PROTEOMICS

ADVANCES IN RESEARCH AND APPLICATIONS

SYSTEMS BIOLOGY – THEORY, TECHNIQUES AND APPLICATIONS

Additional books and e-books in this series can be found on Nova's website under the Series tab.

SYSTEMS BIOLOGY – THEORY, TECHNIQUES AND APPLICATIONS

PROTEOMICS

ADVANCES IN RESEARCH AND APPLICATIONS

RICARDO PARKER
EDITOR

Library of Congress Cataloging-in-Publication Data

Names: Parker, Ricardo, editor.
Title: Proteomics : advances in research and applications / Ricardo Parker, editor.
Description: Hauppauge : Nova Science Publishers, [2019] | Series: Systems biology - theory, techniques and applications | Includes bibliographical references and index. |
Identifiers: LCCN 2019043015 (print) | LCCN 2019043016 (ebook) | ISBN 9781536164404 (paperback) | ISBN 9781536164411 (adobe pdf)
Subjects: LCSH: Proteomics.
Classification: LCC QP551 .P75666 2019 (print) | LCC QP551 (ebook) | DDC 572/.6--dc23
LC record available at https://lccn.loc.gov/2019043015
LC ebook record available at https://lccn.loc.gov/2019043016

Published by Nova Science Publishers, Inc. † New York

CONTENTS

PREFACE

This book provides a brief overview of proteomic techniques used in acute leukemia, including the high-throughput reverse phase protein array, which will highlight the importance of measuring proteins.

Following this, the authors discuss the molecular mechanism of oseltamivir and the advantages and disadvantages of using it. Proteomics- and transcriptomics-based alternatives to oseltamivir are described as well.

Additionally, the most recent research studies aiming to identify biomarkers for the early prediction of spontaneous preterm labor using mass spectrometry-based proteomics are reviewed.

In order to understand the mechanisms of efficient degradation of lignocellulosic biomass, a comprehensive analysis of carbohydrate-degrading and metabolic enzymes is required. As such, the authors summarize a proteome analysis of C. cellulovorans cultured with different carbon sources, and provide some insights as to how C. cellulovorans optimizes diverse enzymes depending on carbon sources.

The concluding chapter focuses on comparative fluorescence gel electrophoresis, which was invented in 2009 and has since then become the prime method for reproducible coordinate assignment in two-dimensional protein polyacrylamide gel electrophoresis.

Chapter 1 - Acute myeloid leukemia (AML) occurs across all ages and although survival is markedly better in the pediatric population, it remains a leading cause of childhood and adult cancer death. AML is characterized by a clonal expansion of myeloid stem cells whose unlimited proliferation properties lead to leukemic cell infiltration of the bone marrow, peripheral blood and solid organs. However, vast biologic heterogeneity of AML is recognized and prognosis varies widely among patients. Current risk stratification is based on biologic features such as molecular genetic abnormalities and cytogenetics. Despite advances in refining individual prognosis that guide treatment strategies, targeting gene mutations and chromosomal translocations remains difficult. Since, the cell net effect of genetic abnormalities, as well as epigenetic and environmental influences, is predominantly determined by protein expression levels and their activation state, assessment of the proteome could potentially contribute to risk stratification and could provide leads for targeted therapy guidance in AML. In this chapter, the authors will provide a brief overview of proteomic techniques used in acute leukemia including the high-throughput reverse phase protein array (RPPA), will highlight the importance of measuring proteins, in addition to genomics, and will discuss the applications of proteomics including the unraveling of biological heterogeneities in the pathophysiology of AML, patient risk classification, protein target discovery and therapy selection. The authors will end with a remark on current challenges and future potential of proteomics in the clinic.

Chapter 2 - The influenza A virus is one of the most dangerous pathogens threatening human life. One of the default treatments for influenza A virus is prescribing the drug oseltamivir. However, due to the mutation rate of influenza A as an RNA-based virus, there are tendencies of resistance toward oseltamivir. In this respect, novel approaches should be devised to ward off the menace of influenza A, mainly with modern proteomics- and transcriptomics-based methods. This review will first discuss the molecular mechanism of oseltamivir and the advantages and disadvantages of using it. Proteomics- and transcriptomics-based alternatives to oseltamivir will be described as well. The authors' proteomics-based computational approach has successfully produced some fine-grained designs, such as cyclic peptide-based, flavonoid-based, and amantadine-derivative leads. The focus going forward should be on the promising design of cyclic peptide because of its stability in our physiological system. At the same time, the availability of natural products based on the biodiversity database and the derivative products of the existing drug should not be overlooked. However, another approach is being developed as well, based upon the transcriptomics method, where, the coding and non-coding RNAs are utilized as drug candidates. The most common drug candidates in use are silencing (si)RNA, which is a short-strand, non-coding RNA. The issues plaguing RNA-based drugs are mainly the delivery mode to the target, as this molecule tends to be unstable in the cell. In this regard, the best approach always depends on the diagnostic results drawn by medical doctors. Despite all of this, oseltamivir remains the default medication for influenza A, in combination with other drugs such as zanamivir.

Chapter 3 - Preterm labor (PTL), defined as delivery <37 weeks of gestation, is a major health issue for modern obstetrics. Approximately 15 million infants are born prematurely worldwide every year and out of those more than 1 million do not reach the second year of life. Preterm infants are also at high risk of long-term adverse effects. The etiology remains unclear but it is likely multifactorial, linked to impaired placental function. Recent studies focusing on placental and maternal peripheral blood protein profiling using mass spectrometry-based proteomic technology show differential expression between normal and complicated pregnancies, providing valuable information about the pathophysiological role of proteins and identifying potential biomarkers for monitoring pregnancy complications. Herein the authors review the most recent research studies aiming to identify biomarkers for the early prediction of sPTL using mass spectrometry-based proteomics.

Chapter 4 - Lignocellulosic biomass has gained much attention as an alternative and renewable carbon source. Lignocellulosic biomass has a highly complex and robust structure, being made up of cellulose, hemicellulose, pectin and lignin. Therefore, the degradation of lignocellulosic biomass to simple sugars is difficult due to its structure. For degradation of such lignocellulosic biomass, an anaerobic bacterium *Clostridium cellulovorans* has attractive features. *C. cellulovorans* secretes a large enzyme complex called "cellulosome" and this complex can degrade lignocellulosic biomass efficiently. For more efficient degradation of lignocellulosic biomass, *C. cellulovorans* optimizes profiles of cellulosomal, and non-cellulosomal enzymes depending on carbon sources. In order to understand the mechanisms of efficient degradation of lignocellulosic biomass, comprehensive analysis of carbohydrate-degrading and metabolic enzymes is required. In this chapter, the authors summarize proteome analysis of *C. cellulovorans* cultured with different carbon sources, and provide some insights how *C. cellulovorans* optimizes diverse enzymes depending on carbon sources.

Chapter 5 - Comparative fluorescence gel electrophoresis (CoFGE) was invented in 2009 and has since then become the prime method for reproducible coordinate assignment in two-dimensional protein polyacrylamide gel electrophoresis. The method is based on the use of at least two fluorescent dyes in one gel run: one for the analyte and one for an internal protein marker, which generates a net of reference spots across the gel. An additional dye can be used to label a standard dilution. Although the marker originally was intended solely for the purpose of providing anchor spots for the correction of analyte coordinates, it may be additionally used for quantification as the fluorescence intensity can be related to the protein amount causing it. The reference net is formed by 8 - 10 proteins ranging from 8 - 100 kDa and 40 wells for marker application so that several standard concentrations can be applied in one run. The major drawbacks of gel electrophoresis – reproducibility and lack of standardisation – thus finally vanish.

In: Proteomics
Editor: Ricardo Parker
ISBN: 978-1-53616-440-4

Chapter 1

PROTEOMICS IN ACUTE MYELOID LEUKEMIA: GUIDANCE TO TARGETED THERAPY SELECTION AND RISK STRATIFICATION

Anneke D. van Dijk[1,†], Fieke W. Hoff[1,†] and Steven M. Kornblau[2,*]
[1]Department of Pediatric Oncology/Hematology,
Beatrix Children's Hospital, University Medical Center Groningen,
University of Groningen, Groningen, The Netherlands
[2]Department of Leukemia, The University of Texas M. D. Anderson Cancer Center, Houston, TX, US

ABSTRACT

Acute myeloid leukemia (AML) occurs across all ages and although survival is markedly better in the pediatric population, it remains a leading cause of childhood and adult cancer death. AML is characterized by a

---

[†] Contributed equally to the work.
[*] Correspondending Author's E-mail: skornblau@mdanderson.org.

clonal expansion of myeloid stem cells whose unlimited proliferation properties lead to leukemic cell infiltration of the bone marrow, peripheral blood and solid organs. However, vast biologic heterogeneity of AML is recognized and prognosis varies widely among patients. Current risk stratification is based on biologic features such as molecular genetic abnormalities and cytogenetics. Despite advances in refining individual prognosis that guide treatment strategies, targeting gene mutations and chromosomal translocations remains difficult. Since, the cell net effect of genetic abnormalities, as well as epigenetic and environmental influences, is predominantly determined by protein expression levels and their activation state, assessment of the proteome could potentially contribute to risk stratification and could provide leads for targeted therapy guidance in AML. In this chapter, we will provide a brief overview of proteomic techniques used in acute leukemia including the high-throughput reverse phase protein array (RPPA), will highlight the importance of measuring proteins, in addition to genomics, and will discuss the applications of proteomics including the unraveling of biological heterogeneities in the pathophysiology of AML, patient risk classification, protein target discovery and therapy selection. We will end with a remark on current challenges and future potential of proteomics in the clinic.

1. Acute Myeloid Leukemia

The acute onset of myeloid leukemia is a malignancy characterized by clonal expansion of poorly differentiated myeloid progenitor cells ('blasts') in the bone marrow. Unlimited proliferation properties of these immature cells lead to leukemic cell infiltration of the bone marrow, peripheral blood and solid organs. Generally, the presence of more than 20% blasts in the bone marrow or blood is required for the diagnosis of acute myeloid leukemia (AML). Median age of onset is 67 years, and although it is a relatively rare disease accounting for only 1% of all cancers, AML occurs across all ages and remains a leading cause of childhood and adult cancer death [1].

1.1. Pathophysiology

Normal hematopoietic stem cells (HSC) are homing in the bone marrow and give rise to all our circulating blood cells. They have the unique ability to self-renew or to differentiate into a multipotent progenitor that eventually differentiates toward a common lymphoid progenitor, leading to the production of lymphoid blood cells, or toward common myeloid progenitors that give rise to the myeloid blood lineage. Lineage specific development is associated with cytokine appearance and protein expression on the cell surface, such as receptors or adhesion molecules. Additionally, lineage-specific gene transcription and translation signatures are recognized [2, 3]. Given that cellular functions are ultimately defined by the post-translational protein state rather than the transcriptome, characterization of lineage- and phenotype-specific protein expression profiles is key for a better understanding of normal and malignant hematopoiesis [4, 5].

1.2. Etiology

The etiology of most AML cases is idiopathic, but known risk factors include genetic predisposition, environmental and chemical exposure, prior hematologic disorders and age. Emerging modern sequencing methods have revealed that AML development is driven by mutations and gene translocations, [6, 7] and given the accumulation of mutations during lifespan, it is not surprising that AML incidence significantly increases with advancing age.

1.3. Risk Classification Based on Cytogenetics

The overall survival for AML patients is ~5-15% in patients older than 60 years and ~40% in the adult population below the age of 60 [1]. In comparison, relatively better outcomes are observed in pediatric patients with ~70% of children that survive [8]. Nevertheless, due to the complexity and heterogeneity of the disease, prognosis varies widely among patients across all ages. Currently AML is classified based on the 2016 World Health Organization (WHO) which classifies AML on the molecular genetic (i.e., mutations) and cytogenetic (i.e., chromosomal abnormalities) features [9].

The latter remains the most important risk classifier in AML and based on this, personalized therapy is expanding. Based on their cytogenetic profile, patients are stratified into a favorable, intermediate or unfavorable risk group [1, 10]. However, about half of the AML patients (e.g., normal karyotype) fall in the intermediate risk characterized group and prognosis in this group is hard to predict underlining the vast biologic heterogeneity of AML [11].

1.4. Risk Classification Based on Mutations

Additional risk stratification in AML is provided by detection of one or more recurrent genetic mutations [7]. Prognostic significance of individual mutations is often determined by the presence of other driver mutations. For example, the most frequent mutated gene is the nucleophosmin *NPM1)* comprising 35% of all AML patients. *NPM1* mutations are predominantly present in normal karyotype AML and in co-occurrence with the fms-like tyrosine kinase 3 (*FLT3*) mutation and yield prognostic significance as *FLT3* mutated intermediate categorized patients have better outcomes with the *NPM1* mutations then those without the *NPM1* mutant [12]. Mutations in tumor suppressor genes (*TP53)* and genes related to chromatin modification (e.g., *DNMT3A*, *IDH1*, *IDH2* or *TET2*) or the spliceosome (*RUNX1, ASXL1*) are also widely recognized [1, 13]. Although ~86% of AML patients have at least two driver mutations, [7] the number of recognized recurrent mutations in AML remain limited compared to other adult cancers as recognized by The Cancer Genome Atlas, and is yet not completely sufficient to explain its heterogeneity [6]. This indeed indicates the existence of other contributing factors beyond genomics.

1.5. Risk Classification Based on Protein Expression

Current risk stratification in AML is predominantly based on biologic features such as molecular genetic abnormalities and cytogenetics in combination with clinical risk factors, with age, performance status, and a history of an antecedent hematological disorder being the most significant. Despite the advances in refining individual prognosis by cytogenetics and genomics that guide treatment strategies, targeting gene mutations and chromosomal translocations remains difficult. It was 21 years between the discovery that the *FLT3-ITD* alternation was frequent in AML and the FDA approval of a drug (Midostaurin) that targets *FLT3* [14, 15] Compared to advances made in all other types of leukemia, it is generally acknowledged that development and implementation of new therapeutic strategies in AML have been slower to arrive, with recent innovations following on the recent increases in our knowledge of AML biology and etiology [16]. However, despite the recent identification of a set of genes that are commonly mutated genes in AML, targeting these has been challenging, as most have proved non-druggable to date (e.g., *NPM1*), with the drugs targeting only the mutant forms of *IDH1* and *IDH2* as a notable exception. Since the protein products of these mutated genes function as proteins, in the context of networks of interaction with other proteins, it may be possible to target other members of the interactome, if the mutated gene or protein cannot be targeted. But to do so requires knowledge of activation states, not just expression levels. We here argue that the comprehensive study of protein expression, including post-translational modifications (PTM), so-called proteomics, may better bridge the gap between AML and its heterogeneous landscape. Proteome assessment could result in better risk stratification and could potentially provide leads for the selection of targeted therapy in AML.

While it has long been hoped that mRNA expression would predictably and reliably correlate with protein levels, combining high-throughput genomic and proteome approaches have taught us that the Pearson correlation coefficient only varies between 0.17 and 0.40 when mRNA is correlated with protein levels [17-19]. This poor correlation between mRNA transcription and protein abundance is more recently confirmed in colon- and rectal cancer, [20, 21] gastric cancer, [22] breast cancer, [23] and medulloblastoma [24]. In our own AML data we only observed a correlation of 0.17 (unpublished data). It is therefore clear that the basic dogma describing that mutations in the DNA leads to changes at the mRNA expression level and results in quantitative or qualitative changes in protein function is too simplistic and hinders the translation of genomic discoveries into useful therapeutics.

This lack of correlation is not surprising as after transcription mRNA transcripts undergo alternative splicing and mRNA editing which increases the total potential of ~25.000 genes into ~100.000 different transcripts that can be translated into proteins. Furthermore, the kinetics of how rapidly and how often a mRNA molecule is translated can vary and can be affected by miRNAs targeting the transcript as well. The additional heterogeneity in the longevity of a protein molecule further decreases the correlation between mRNA levels and protein levels. Subsequent PTMs, including phosphorylation, acetylation, methylation, and ubiquitination, also increases the proteome complexity [25]. PTMs may change protein folding, localization, lifespan, function and therefore its activation state, which is unmeasurable via genomics. Also, an understanding about the protein-protein and protein-DNA interactions and how different protein complex assemblies are built is crucial for our understanding of cellular behavior. As proteins are affected by genetics, chromatin remodeling, micro-environmental influences and their PTMs, the potential raises the number of 100,000 differentially transcribed proteins into more than a million potential variants. Proteomics may be the least developed and investigated '-omics' approach, yet it is likely the most important and informative.

2. Methodologies to Study Proteomics

2.1. High-Throughput Proteomic Techniques Used in AML

Proteomics is the large-scale study of the proteome which, in theory, covers the entire set of expressed proteins and their PTMs in a particular cell or organism. In practicality with available methods, most studies only provide information on a large number of proteins and not the whole potential proteome. Numerous methods are available to study proteomics, but globally two classes of high-throughput techniques are used in AML research, which will be discussed here; mass-spectrometry (MS)-based techniques and antibody-based techniques.

2.1.1. MS-Based

MS is a powerful tool that allows qualitative and quantitative assessment of proteins present in highly complex samples. The analysis starts with the formation of ions (charged fragments) from the protein analyte, which are subsequently sorted using electrical and/or magnetic fields based on their mass-to-charge ratio (m/z). An ion detector enables determination of the intensity of each separated m/z fraction and based on the abundance of the fragments enables identification of the protein [26]. There are two main methods used for the ionization of a protein: matrix assisted laser desorption/ionization (MALDI) and electrospray ionization (ESI). In MALDI the protein sample is mixed with an energy absorbing matrix [27]. Laser energy is used to irradiate the matrix which causes vaporization of the matrix together with the sample, resulting in the formation of ions. Since MALDI mainly forms large fragments with minimal fragmentation, it is most often combined with a time-of-flight (TOF) mass-spectrometer, which measures the travel time from an ion to reach the detector, meaning that the lighter the fragment, the quicker the ion travels [28]. A variation of this method is the surface-enhanced laser desorption/ionization (SELDI). SELDI is similar to MALDI, but the difference is that the proteins in the sample are bound to a surface instead of being mixed with the matri. [29].

A second way to ionize a protein is with the use of ESI [30]. ESI creates ions using electrospray to dissolve the protein lysate. High-voltage is applied to the dissolvent to create an aerosol of small charged fragments. When a protein sample is highly complex, samples may require separation prior to MS analysis using 1D or 2D gel electrophoresis, high-pressure liquid chromatography (LC-MS), or gas chromatography (GC-MS) to maximize the sensitivity.

A study by Nicolas et al. used SELDI-TOF MS to characterize 54 de novo AML samples. Based on the affinity of the surface binding proteins, they identified two proteomic signatures that were significantly associated with a difference in overall survival and disease-free survival. Within these two signatures, they found S100A8 as the protein with the highest discriminative value and suggested this protein as a potential marker to predict inferior survival [31]. Another study by Xu et al. used SELDI-TOF MS to analyze 189 AML samples including different AML subtypes (e.g., AML-monocytic, AML-granulocytic, acute promyelocytic leukemia (APL)) and samples from healthy controls. They developed a proteomic-based classification approach capable of replicating the morphological and differentiation-based classification scheme of the well-established French-American-British system, but they did not report further on the differences between the protein signatures, concluding that this can potentially serve as new diagnostic approach [32].

2.1.1.1. Proteomics of the Post-Translational Modification

Protein activity is regulated by the PTMs, including phosphorylation, glycosylation, ubiquitination, oxidation, and cleavage. This makes the knowledge of the expression of a particular protein alone insufficient to define the activity of a particular portion, the activation state of a protein must also be known. As one of the major challenges of MS is the depth of the protein coverage, which depends on the dynamic range of protein concentrations in a human cell, this complicates the identification and quantification of proteins with a low abundance, and in particular of the PTM. Traditionally, a 'bottom-up' or 'shotgun approach' has been used to study the PTM, which uses pre-separation and pre-digestion steps, to increase the yield of small protein coverage sequences. However, the limitations of this approach are that 1) due to the complexity of the sample, only a small fraction of the PTM are detected, 2) the redundancy of small fragments makes it difficult to reliably determine which peptide is a product of which protein, and 3) the loss of information of the PTM, as the localization process of the modification site to one or more of often several plausible amino acids is difficult due to the overload of small fragment out of the context of the total protein.

To overcome those limitations, one solution is to perform enrichment steps after digestion of the protein lysate to increase the number of identified PTM. The most extensively studied PTM is probably phosphorylation, and the most preferred method to study this is with the use of metal ion affinity chromatography (IMAC). This technique allows phosphoproteins to be retained in a column containing immobilized metal ions based on the affinity of a phosphate group to metal ions, which results in a higher abundance of the protein of interest in the sample [33]. Another method to study the PTM is with the 'selected reaction monitoring' (SRM) approach. SRM filters peptides with a particular mass-to-charge ratio (precursor ion) and secondly activates those ions to generate compound-specific product ions which are then filtered and detected. This makes it possible to distinguishing highly similar proteoforms such as isoforms and PTM proteins, including phosphorylation and ubiquitination. Because this method only analyzes samples with a particular mass-to-charge, it can detect low abundance proteoforms, which are below the limit of detection of standard MS analyses [34]. Utilizing this technique in AML, Matondo et al. studied the effect of proteasome inhibitors on the proteome of two different AML cell lines and identified more than 7000 proteins. Based on the results from this shotgun approach, several candidate proteins that were present with a low abundance were then further validated using SRM in the same two cell lines [35].

2.1.1.2. Quantitative Proteomics

MS can provide information on the quantity of a protein. Globally, there are two methods that can be used in the MS-based workflow; label-free and stable isotope labeling. Label-free quantification does not use any protein or peptide labeling, but counts the peak of the peptide intensities or uses spectral counting [36]. This approach may be less accurate compared to isotope labeling, which uses metabolic labeling for instance by stable isotope labeling with amino acids in cell culture (SILAC) or chemical labeling using isobaric tags for relative and absolute quantitation (iTRAQ) or tandem mass tag (TMT) [37, 38]. Sandow et al. used SILAC-labeling to quantitatively study the effect of EZH2 activity inhibition on AML induced in mice. They identified proteins involved in nucleic acid binding (e.g., TOP2A) and HSC differentiation (e.g., FCGR2 and ITGAM) that were significantly increased or decreased after treatment with epigenetic therapy [39].

2.1.2. Antibody-Based

Another approach to study proteomics in AML is the antibody-based proteomic approach. In general, there are two types of antibody-based methods; tissue microarrays (TMA) and protein microarrays (PMA). TMA consists of paraffin blocks that contain up to 1000 samples that allows the study of multiple analytes (e.g., bone marrow biopsies) in a single experiment. TMA is often used in combination with immune-histochemistry (IHC) to identify protein expression patterns and to study and validate protein biomarkers. For instance, TMA in combination with IHC was used to study the expression of CD74 in 248 AML cases, which was subsequently correlated with clinical parameters in AML [40]. Mattsson et al. studied protein expression levels of *p53*, *p21*, *p16*, and *PTEN* before and after hematopoietic stem cell transplantation in bone marrow samples from 34 pediatric AML patients using IHC. They found significant difference in *p53* with overexpression in the group of patients who relapsed compared to the relapse-free patients at >3-6 months post-hematopoietic stem cell transplantation [41].

PMA can be further divided into forward phase protein arrays (FPPA) and reverse phase protein arrays (RPPA). FPPA used a platform on which antibodies are immobilized on the array with known positions. Samples are then printed on the array and in the presence of a particular protein bind to the antibody. After exposure to a secondary antibody, protein expression intensities can be measured. Each slide is exposed to a single protein sample, and for each protein sample expression of multiple proteins can be measured simultaneously. RPPA uses the 'reverse' approach, printing the samples on the slide and then each slide is probed with a single detection antibody followed by a secondary antibody to amplify the signal.

In comparison to MS-based methods, the disadvantage of RPPA is that it cannot be used as a *de novo* discovery platform. In theory MS can be used to identify all proteins although in practice this requires depletion of highly abundant proteins, to enable the detection of low abundance proteins. RPPA has some advantage over MS: 1) it requires less material (approximately 3 x 10^5 cells to test 400 different proteins), which makes it highly suitable for clinical applications; 2) it analyzes all samples at once, allowing a direct comparison of protein expression across samples, a process that is much more difficult in MS where each sample is studied individually at multiple timepoints; and 3) it is far more economical when studying a large number of samples. In particular its potential to analyze large sets of patients simultaneously is very important, as analyzing small groups of patients may not fully represent the heterogeneous AML population and thus may miss important protein utilizations in subgroups of AML. Contrary, disadvantages of RPPA are that 1) all samples must be printed at one time, making it impractical for real-time individual patient analysis; and 2) RPPA is biased to proteins and isoforms for which a strictly validated antibody is available. As RPPA is a high-throughput antibody-based technique, the reliability is highly dependent on the quality of the antibodies that are used. Unlike in a Western blot, with RPPA there is no separation of the proteins according to molecular weight, which has the consequence of making signals from potential cross-reactivities indistinguishable from the intended signals. Antibodies used on RPPA must all be strictly validated, and must be shown to be specific, selective, and reproducible in the context for which it is to be used, must perform robustly across different sample types and must acts consistently over time. However, once an antibody is validated, RPPA allows not only detection of total proteins, but can also yield information about PTMs since a broad spectrum of antibodies that detect phosphorylated, acetylated, methylated, and cleaved forms of proteins is available.

3. RPPA Cell Lysate Selection and Preparation

3.1. Studying the Right Cell Populations in AML

Our group is using RPPA to study leukemia in adult and pediatric populations diagnosed with AML, APL, acute lymphocytic leukemia (ALL) as well as in mesenchymal stromal cells. The major advantage of studying leukemia is that primary leukemic samples can be easily isolated from peripheral blood samples or bone marrow aspirates that are obtained during diagnostic procedures without performing additional bone marrows on patients. However, in order to obtain a representative analysis of the leukemic cells, purification of the leukemia cells from other cells is crucial. As cell surface markers are well defined for the leukemic blast, this makes it easy to purify or enrich the sample from the contaminating non-leukemic cell. For instance, ficoll separation can be applied to separate the erythrocytes and neutrophils based on their density. Fluorescent-Activated Cell Sorting (FACS) or Magnetic-Activated Cell Sorting (MACS) can be used to remove the contaminating B- and T-cells from the myeloid blast cells based on the presence of the cell surface marker CD3 on the T-cell and CD19 on the B-cell. However, as patient material is often processed quickly after collection, before the phenotype of the AML is known, this makes depletion of the non-malignant macrophages, monocytes and dendritic cells that may be present in the AML samples (<10%) impossible as this would remove leukemic blasts in AML with a monocytic lineage (approximately 30-40% of the cases are myleomonocytic = FAB M4 or monocytic = FAB M5 cases).

RPPA is also suitable to additional comparisons based on location or disease state. Information can be gained from studying the proteome in different cell types or different subcellular localizations (e.g., nucleus, cytoplasm, mitochondria). We have recently generated a RPPA with leukemia cell lines and patient samples that were so fractionated (unpublished data). Likewise, since RPPA samples are batched before printing it is possible to collect samples at multiple time points for comparison of changes over time, including samples obtained at time of first diagnosis, obtained from patients that are primary resistant, or from patients diagnosed with relapse AML. Since relapse or primary refractory AML are thought to arise from inherently chemoresistant leukemia stem-like cells (LSC) which have self-renewal capacity, studying the proteomics of the LSC may be important to understand the pathophysiology of AML, especially if compared to the normal hematopoietic stem cell (HSC) or the leukemic blast. However, as cell markers that differentiate LSC may differ between patients, and as LSC only occur with a frequency of 1 in 10,000 to 1 in 5 million in AML, this makes the isolation of enough cells for experimental analyses challenging [42, 43].

However, the cell sparing requirements of RPPA enabled us to isolate and study cell subsets that were enriched for stem cells ($CD34^+$, $CD34^+/CD38^-$) and demonstrated that the protein expression of LCS was markedly different from bulk leukemia and $CD34^+$ cells [44].

3.2. AML Cells Collected from Peripheral Blood or Bone Marrow

Another consideration that was previously examined by our group is whether the source of the AML blast matters. In that study, paired peripheral blood and bone marrow samples were collected and analyzed the RPPA for 230 different antibodies against proteins involved in many different signal transduction pathways [45]. In general, no global differences in expression were found for the AML samples. A previously study showed similar results, with only 25/176 antibodies that showed different ($p \leq 0.01$) expression levels between the compartments, with 10 (6%) being higher in the blood and 15 (9%) higher in the marrow, closely mirroring expectations for that statistical cutoff. For the majority of these the fold difference was <25%, so whether these statistically significant differences have biological differences are questionable. Likewise, Braoudaki et al. performed MALDI-TOF on 5 BM and 5 PB samples from AML and found that the protein patterns obtained from the analyses of AML plasma samples did not exhibit significant disparities between BM and PB specimens for most proteins [46]. However, a critical caveat is that samples obtained for research are usually collected as the 4th or subsequent pull as part of a procedure being done for clinical purposes, and there is some unknowable degree of hemo-dilution that may blur distinctions between blood and marrow. So, for practical purposes BM and PB produce similar results and can probably be used interchangeably.

3.3. Impact of Freshly Prepared Versus Cryopreserved Samples on Protein Quality

Protein lysates can be prepared from samples on the day of arrival in the lab ('fresh') as well as from cells that were first cryopreserved ('frozen') and stored and then prepared at a later time. However, we previously noticed that when we made protein preparations from rapidly thawed cells, without selection, we observed markedly higher levels of cleaved caspases and PARP, likely reflecting the apoptosis and cell death that occurred with the freeze-thaw cycle. To prevent this artifact, we adopted a protocol of thawing cells and then rapidly putting them into 37°C media with 20% FCS in an incubator for 2-4 hours for stabilization, followed by ficoll separation to remove dead and dying cells before preparing the protein lysate. With this additional handling and time in culture we no longer observed the differences in cleaved caspases and PARP. Although we initially thought that this made it possible to use fresh and frozen samples interchangeably, [47] we recently showed that paired samples, that were either freshly prepared or frozen, were still significant altered in protein expression [45]. Out of the 228 proteins, 162 (71%) proteins showed significant differences in expression between fresh and frozen samples (two-sided t-test, $P < 0.05$ corrected for multiple testing). This result is not surprising given the additional handling, time in culture, and stimulation by chemokines and cytokines present in the FCS in the media. Therefore, we are convinced that RPPA should be restricted to identically processed cells, either all fresh, or all cryopreserved, to remove the batch effects due to cryopreservation. However, with the caveat that differences have been induced by processing when cryopreserved. Also, we have demonstrated that samples must be kept on ice until processed, and must be processed within 72 hours of collection to avoid additional handling induced changes (unpublished data).

4. Utilizing RPPA to Study Protein Expression in Leukemia

4.1. Use of RPPA to Study Single Proteins

Once a RPPA has been generated and stained with a list of antibodies the data for each individual protein can be analyzed separately or collectively. In adult AML, individually studied protein have included: the Absent, Small, Or Homeotic Discs-like Protein (*ASH2L*), [48] Friend Leukemia Virus Integration 1 (*FLI1*), [49] Forkhead O Transcription Factor 3A (*FOXO3A*), [50] Galectin 3 (*LGALS3*), [51] Transglutaminase 2, [52] and Tripartite Motif Containing 62 (*TRIM62*) [53]. For each of these, comparisons can be made to clinical, genetic and molecular events as well as to response and outcome. For instance, *FLI1* expression in AML patients was higher or equal compared to normal CD34^{+} cells in 32% and 5% of the patients respectively, and patients with low or high *FLI1* had a shorted remission duration and overall survival. For *ASH2L*, lower expression levels in AML were associated with increased overall survival.

RPPA enables us to study PTMs of individual proteins that reflects the functional state of the cancer cell. The relevance of studying PTMs in AML is shown in our report about *FOXO3A* [50]. *FOXO3A* acts as tumor suppressor in many cancers by regulating transcription of apoptotic associated genes. However, converse roles of *FOXO3A* expression are described in leukemia and activity status is dependent on nuclear localization. As transport of *FOXO3A* is guided by protein modification, i.e., phosphorylation, functional assessment of *FOXO3A* should only be performed focusing on phosphorylated *FOXO3A* rather than on total levels. For instance, phosphorylation of *FOXO3A* on serine 256 or 318 by *AKT* inhibits transcriptional activity by enhancing *FOXO3A* nuclear export resulting in increased cell proliferation. Less total expression as well as more phosphorylation thus might decrease *FOXO3A* function. To functionally assess *FOXO3A* protein expression in leukemia, we measured *FOXO3A* total protein expression and phosphorylation on serine 318 in 511 newly diagnosed AML patients using RPPA. Both total and phosphorylated *FOXO3A* were lower expressed in the AML patient samples compared to normal $CD34^+$ cells. Further analysis was performed using the phosphorylation/total expression ratio (*PT-FOXO3A*). *FOXO3A* is known to regulate transcription of apoptotic proteins, such as *BAX*, *BAK* and *BAD* and as expected, *PT-FOXO3A* significantly correlated with these proteins on the same RPPA array. Most interestingly, relative high phosphorylation levels (i.e., high *PT-FOXO3A*) were highly prognostic among patients, but not total F*OXO3A* levels. Patients within highest tertile of *PT-FOXO3A* expression had adverse outcomes in terms of overall survival. This was also true in cytogenetically normal and unfavorable patients. *PT-FOXO3* yielded prognostic significance in multivariate analysis showing that the *FOXO3A* PT-ratio is an independent predictor of outcome in adult AML. Moreover, the lifespan of *FOXO3A* is regulated by ubiquitination and subsequent proteasomal degradation and acetylation of *FOXO3A* has been observed to downregulate *FOXO3A* activity. We here only analyzed phosphorylation of *FOXO3A* on serine 318, but a more comprehensive analysis of other phosphorylation sites, acetylation and ubiquitination of *FOXO3A* might further improve prognostic discrimination.

4.2. Identification of Protein Expression Patterns Based on the Global Proteome

We next wanted to see if there were recurrent patterns of expression (constellations) across the studied cohort encompassing the entire studied proteome. This shift away from studying individual proteins was supported by the assumption that the net effect of protein expression and protein activation is determined by the combined influences of all proteins in a cell, rather than on changes in a single protein. Our initial approach was to perform unbiased hierarchical clustering as reported in our initial AML RPPA with 256 newly diagnosed AML patient samples with different recurrent cytogenetic abnormalities, for 51 total and phosphoproteins [54]. Proteins were involved in a wide range of cellular processes. Expression data was clustered based on the absolute value of the Pearson's correlation coefficient and this identified 10 groups of strongly correlated proteins, in which many proteins had related functions supporting the idea that the approach has functional validity. From here, mathematical analysis separated patients into 7 protein signatures. Protein signatures provided prognostic information for overall survival, remission attainment and relapse rate, distinct from cytogenetics. Because mesenchymal stromal cells (MSC) support AML cell survival, this approach was also applied to RPPA expression data from 28 proteins that were differentially expressed between MSC cells from AML patients (n = 106) and healthy controls (n = 71) [55]. This identified 3 constellations of strongly correlated proteins that made biologically sense, allowing division of samples into four clusters that were significantly associated with survival.

While this confirmed our original hypothesis that there is a finite number of recurrent signatures, we felt that this approach might be less than optimal. Due to some inherent considerations of how hierarchical clustering (HC) is performed. First, HC weighs all proteins equally, in all situations across the dataset, but a protein may be relevant in one signature, but irrelevant in another. Second, HC is agnostic to all known functional relationships between proteins, and ignoring known interactions is intuitively simplistic. Third, HC requires that all pieces of data be considered and placed into a group, even if it does not fit well (if one sorts a deck of cards into suits where do you place the two jokers?). This inability to separate informative proteomic wheat from non-informative chaff degrades the quality of the clustering. We therefore set out to come up with a computational methodology that dealt with the limitations of unbiased HC.

This methodology first allocates proteins into groups of proteins with a related function based on existing knowledge or strong association within this dataset (Protein Functional Group (PFG)) (Figure 1). Within each PFG protein clusters can be identified, based on the strength of correlation between protein expression and activation status of the core protein members of that PFG within a group of patients. Individual PFG and their associated clusters can be analyzed individually, as discussed below for

TP53 [56] and histone modifying proteins, [57] or collectively as discussed below in the 'MetaGalaxy' section.

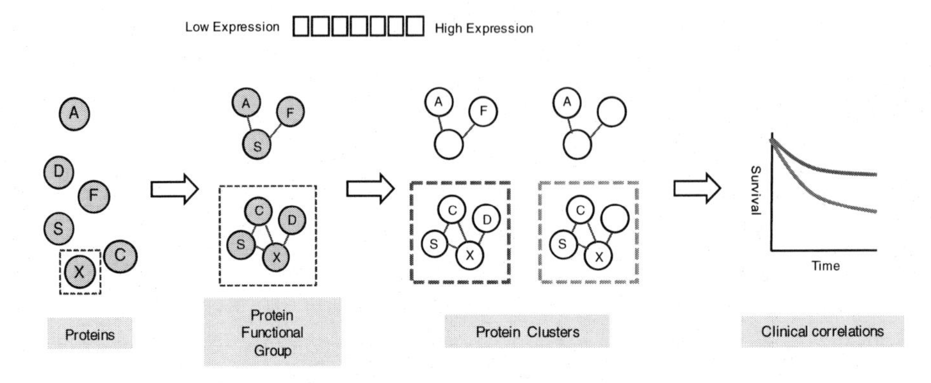

Figure 1. Protein functional group analysis.

This figure represents a schematic overview of the protein functional group analysis. Proteins are first divided into groups of functionally related proteins (e.g., *TP53*, histone modification, cell cycle), based on the existing literature or based on strong correlations within the data set. Clustering algorithm was applied to identify an optimal number of protein clusters. Protein clusters are defined as a group of patients that expressed similar (correlated) expression of core protein members (shown in the red and pink box). Finally, protein clusters were correlated with clinical characteristics and disease features.

4.2.1. Protein Expression Patterns That Modulate P53

In half of human cancers, the tumor suppressor gene *TP53* encoding for *p53* is mutated inducing protein stabilization and affecting the transcriptional regulatory activity of *p53*. In AML, *TP53* mutations are found in 5-8% of newly diagnosed patients, and in 30-40% of therapy-related AML, and are recognized as a poor independent risk factor in AML. In contrast to the *TP53* genetic mutation, less is investigated about the prognostic value of *p53* protein expression in AML patients. Our group has recently studied protein expression levels of *TP53* and found that patients with high *p53* protein expression patterns have significantly shorter overall survival and complete remission durations and have higher relapse rates compared to those with low *p53* protein expression [56]. Interestingly, these findings were independent of *TP53* mutational status. This excited us to further evaluate the protein expression patterns of proteins that regulate *p53* status. The *p53* function can be suppressed by the overexpression of its negative canonical regulators, e.g., *MDM2* and *MDM4*. Overexpressing *MDM2* has been described in multiple human cancers and it has been indicated that these cancer cells may bypass the tumor suppressor role of *p53* in wildtype *TP53* conditions. In our study we indeed found *MDM2* and *MDM4* overexpression in patients with high *p53* protein levels and wildtype *TP53* alleles, including significant lower expression of the *p21* effector. These findings suggest that aberrant *p53* expression is triggered by other mechanisms beyond mutations or deletions of the *TP53* gene which may explain poor outcomes of patients with high *p53* protein expression regardless of *TP53* mutational status.

Using RPPA we were able to simultaneously assess protein levels of total *p53* and phosphorylated *p53,* its regulators, e.g., *MDM2, MDM4, TRIM24,* and phosphorylated IRS1 as a *p53* target. In total 9 proteins were gathered into this '*TP53* functionally related group' based on existing knowledge and strong data-driven interactions. Our clustering approach [58] identified 7 as the optimal number of 'protein clusters,' subsets of patients that showed similar expression of the *p53* and associated proteins. These 7 distinct proteomic clusters in addition yielded prognostic information. Patients that had membership in the protein cluster with extremely high *p53* had worst overall survival, accompanied by low *MDM2, MDM4* and *p21.* As in total 223 antibodies were analyzed on this RPPA, the creation of a protein-protein interaction network based on correlation coefficients within the dataset revealed protein networks for each of the identified protein clusters, uniquely defined by individual expression levels. Per example, patients within the poor prognostic cluster that expressed extremely high *p53* protein expression also displayed very high expression of *p53*-regulated anti-apoptotic proteins.

Although *TP53* mutational status is prognostic in many cancers, this study is one example that highlights the importance of examining the actual post-translational protein expression status rather than only the mutational status. *TP53* is infrequently (5-8%) mutated in newly diagnosed AML patients, but *p53* protein expression can already provide prognostic information when it is dysfunctional via unforeseen *MDM2* and *MDM4* overexpression regardless of TP53 deletion or mutation. The usage of the sensitive and high-throughput RPPA may indeed help bridge the gap between AML and its enormous heterogeneity.

4.2.2. Histone Modification Protein Expression Patterns

An important regulator of gene expression that helps to define the phenotype of an individual cell is the epigenome. Epigenetics literally stands for the study that modifies the chromatin structure and regulates gene expression programs without altering the genetic code [59]. DNA methylation, histone tail modification and microRNA expressions are the most studied and well-known forms of the epigenome and modifications are increasingly observed in cancer cells since the emergence of sequencing approaches [60, 61]. Of note, acquired global hypomethylation, local CpG island hypermethylation and an imbalance between histone acetylation and deacetylation are recognized as altered epigenetic programs in cancer, however the genome-wide chromatin modification profile including expression pattern of multiple histone marks remains uncharacterized [62].

In our dataset of available protein antibodies for RPPA, we identified a Histone Modification PFG with 20 proteins that play a pivotal role in histone modification and chaperone processes along with proteins that showed a strong correlation with these in the dataset, hereafter referred to as 'histone' group [57]. Clustering analysis of these 20 proteins in 205 adults with newly diagnosed AML patients revealed an optimal number of 5 clusters. Two clusters formed a proteomic profile with upregulation of the histone group proteins: *KDM1A, NCL, SIRT1, hnRNPK, BRD4* and *NPM1*. Individually these proteins are described in the literature to play a role in leukemia, but RPPA allowed us to assess simultaneous protein expression of these proteins related to histone modification for the first time. Patients with a proteomic profile that showed global high expressions of histone group proteins had a significantly worse overall survival compared to all other clusters and this was independent of cytogenetic status. The high expressing histone proteomic profile was associated with the poor prognostic *FLT3* mutation; however, this pattern was prognostically adverse in patients with intermediate cytogenetics regardless of *FLT3* status. Of note was the observation that all patients in the high expressing cluster who responded to demethylation agents and HDAC inhibitors had the *FLT3* mutation. Since *FLT3* mutants were independent of the outcome assembled in the high expression histone proteomic profile, it shows us the possibility that demethylation agents and HDAC inhibitors may contribute in treatment approaches for AML patients with both the indicated proteomic profile and *FLT3* mutation. Furthermore, the function of histone tail modifications and direct DNA methylations are interdependent on each other, especially with respect to gene silencing [63]. However, we observed no significant relation between mutations affecting DNA methylation and the proteomic profiles identified by RPPA.

This study again highlights 1) the importance of analyzing proteomic signatures instead of a single protein and 2) the prognostic value of proteomics in addition to well-studied genomics in AML. We here present that proteomic profiling of histone modification related proteins reveals a poor prognostic group regardless of molecular genetic status and cytogenetics. Histone modification is a dynamic process and is in charge of regulating the gene expression programs [59]. As genomics on its own are not sufficient to cover the largely identified dynamics of the epigenome, combinational analysis of the genome, transcriptome and proteome will be key to further identify functionally and clinically relevant chromatin profiles in AML.

4.2.3. Recurrent Protein Expression Patterns in AML; The 'MetaGalaxy' Approach

Analysis of an individual PFG, while informative, does not provide information on how the activity between the different PFG relate to each other, or on the entire proteome within a patient. We hypothesized that there would be recurrent patterns of interaction between the various PFG clusters that would form a finite set of protein expression signatures that are shared by different subsets of patients. To recognize these, we developed a novel computational approach in which patients were clustered based on their protein cluster membership which we called the 'MetaGalaxy' approach (Figure 2). For each patient it was denoted whether the expression pattern of a particular PFG was present (set to = 1) or absent (set to = 0) in their leukemia, using a binary matrix system. Block Clustering, [64] a method that allows simultaneous clustering of the rows and columns of a matrix, was used to cluster patients (columns) and their assigned protein cluster memberships (rows). Correlation between protein clusters was defined as a 'Protein Constellation'; a group of protein clusters from various protein functional groups that were strongly correlated with each other. Based on constellation membership of patients, we were able to identify subgroups of patients that expressed similar combinations of protein constellations, which was defined as a 'protein expression signature'. This methodology was already applied to several different RPPA data sets, including data from patient samples from adult AML, adult APL, pediatric AML and pediatric acute lymphoblastic leukemia (ALL) [65-68]. Currently, we are applying the same approach to samples from adult ALL patients, pediatric T-ALL, adult chronic lymphoblastic leukemia (CLL) and to samples from patients diagnosed with the myelodysplastic syndrome (MDS) (manuscripts in preparation).

With this segmented approach a substantial amount of structure was observed across each data set, with an optimal number of protein constellations, ranging from 10-13. Constellations were significantly associated with protein signatures and within each data set we found constellations that were exclusively present in a particular signature, as well as constellations that were present across multiple signatures. This suggests that despite the heterogeneous character of acute leukemia, there are patterns in protein expression that are more homogeneously expressed than others. Signatures were also significantly associated with leukemia subtypes. In pediatric ALL clear distinction was observed between the T-cell–specific signature 1 (100%), the pre-B ALL dominant signatures 2, 3, and 4 (83%) and pre- B-ALL exclusive signatures 5, 6, and 7 (100%). Interestingly, while constellations 3 and 5 were only found in T-cell ALL and constellations 2,

4, 6, 8, and 10 were exclusive to pre-B-cell ALL, constellations 1 and 9 showed some overlap between pre-B cell ALL and T-cell ALL, suggesting shared protein deregulation [68]. When we compared those pediatric ALL samples to pediatric AML samples that were printed on the same RPPA, we again found clear separation in disease, not only within T-ALL and B-ALL, but also between AML and ALL. Only a few patients showed an overlapping protein profile between AML and ALL. Three out of the 10 constellations were present in both diseases. Unpublished data from 205 adult AML and 166 adult ALL patients identified the existence of 11 protein signatures, of which 5 were AML dominant (93-100%), 4 were T-ALL dominant (79-100%) and 2 signatures contained a mixture of AML, B-ALL, T-ALL samples (50% and 68% AML). Three out of the 12 constellations were predominantly associated with AML, 4 were associated with ALL, 2 were associated with a mixture of ALL and AML cases, and 3 signatures were not strongly associated with any particular signature). This again highlights that 1) AML and ALL are completely different diseases, not only in terms treatment, genetics, outcome and epidemiology, but also on protein level given that they can be clearly separated from each other, and 2) that despite this significant separation, particular protein expression patterns are overlapping between both diseases, suggesting that some shared deregulations exist between AML and ALL, which can potentially be clinically relevant.

Signatures were also significantly correlated with clinical outcome and response to therapy (e.g., primary resistance to induction therapy, complete remission, relapse), and in adult AML this stratification generated greater prognostic separation than was observed with clustering based on individual protein expression levels. Even after adjusting for common prognostic factors (i.e., age, cytogenetics and white blood cell count) signatures remained significant independent prognostic factors using the cox proportional hazards regression model. Together, this shows that there are recurrences in proteomic data that may have biological and clinically relevant information that could be used for future applications.

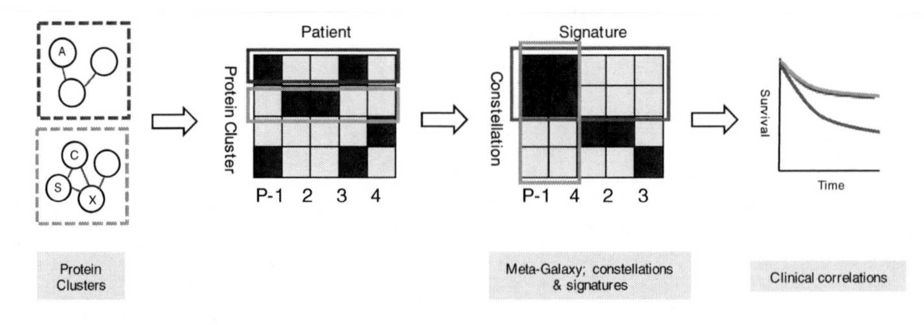

Figure 2. MetaGalaxy analysis.

This figure illustrates a schematic overview of the 'MetaGalaxy' approach. Briefly, protein clusters from the PFG were assembled into a large binary dataset that indicated the protein cluster membership for each patient. Each row represents a single protein cluster. Each column represents an individual patient. For instance, if a patient is a member of Histone Modification protein cluster 1, this patient is coded as a 1 (blue) for this protein cluster 1 and as 0 (yellow) for Histone Modification protein cluster 2, 3, 4 and 5. Block clustering was applied to search for recurrences in protein cluster membership. Protein clusters that tend to co-correlate strongly with each other were defined as a 'protein constellation' (red box, horizontally). A group of patients that expressed a similar pattern of protein constellations was defined as a 'protein signature' (green box, vertically). Protein signatures are correlated with disease outcome and clinical characteristics.

5. Discussion

5.1. Future Application of Proteomics; Translation to the Clinic?

In this chapter we have described the use of proteomics in AML. Over the years, hundreds of proteins have been identified in AML that were associated with disease characteristics, outcome or disease progression, but so far few of them have entered the clinic. Thus, the question remains, how can we bring proteomics to the clinic and what are potential future applications?

5.1.1. Proteomics to Aid Risk Stratification

The current WHO classification incorporates cytogenetics and molecular changes, and most prognostic stratification systems are based on these in association with clinical features, typically dividing patients into favorable, intermediate and unfavorable groups. However, within these groups there is significant heterogeneity of response suggesting that these prognostic classifications could be improved on. Furthermore, despite the usage of mutations in predicting disease progression, few of the most common mutations (*FLT3*, *DNMT3A*, *NPM1*), have no drugs that just targets the mutation, the way the *IDH* inhibitors do (although *FLT3* inhibitors that target wildtype as well as mutant *FLT3* are clinically approved in AML). This limits the ability to convert knowledge of molecular abnormalities into specific therapeutic interventions. As noted above, dysfunctional protein expressions often confer prognostic information besides the existence of mutations. Proteomics could therefore be a powerful means to add and to improve risk stratification.

We hypothesize that adding proteomics can likely help to identify more refined subgroups that better correlate with risk. Clinically, this is important as it can help by the selection of patients that need more aggressive chemotherapy regimens (or stem cell transplantation) as well as patients that can be treated with less toxic regimens. If we can discover a protein or a limited set of proteins in AML as a kind of biomarker that is associated with outcome or disease progression with a high sensitivity and specificity, we can develop a rapid test (e.g., ELISA, IHC, or FPPA) that quickly provides information about the protein expression in the AML patients.

5.2. Identification of Protein Targets for Personalized (Combinational) Treatment Strategies

Many novel drugs now in development target proteins in pathways downstream of mutational events, rather than the actual mutated protein. We hypothesize that proteomics can potentially be used for drug development and therapy selection. Many studies, including our RPPA results, have identified potentially targetable proteins with dysregulated expression levels in subgroups of patients with AML. Once validated in independent and large cohorts, these findings could potentially be clinically useful. Significant over or under expressed proteins may contribute to the ungoverned growth of the leukemia population, likely in combination with other deregulated proteins. Assuming that this is true, then inhibiting overexpression or replacing under expressed proteins (in case of inactivity, or a loss of function) would disrupt the function of this protein within the leukemic cell. Ideally, this could reveal several proteins that could be targeted individually, in combination, or in combination with conventional chemotherapy to improve response.

Among the drugs that are currently in development for AML or are tested in clinical trials, those altering the apoptosis regulation are one of the most promising so far. For instance, venetoclax (formerly ABT-199) is a *Bcl-2*-selective inhibitor. *Bcl-2* is an anti-apoptotic protein of the *BH3*-family that is involved in the intrinsic (mitochondrial) pathway of apoptosis and is frequently upregulated in AML. As venetoclax is only effective in some patients, it is important to predict who will benefit and who will not benefit to avoid unnecessary exposure and to optimally use its potential. To investigate this, Souers et al. has shown that sensitivity to *Bcl-2* inhibitors was exhibited in some cell line panels that correlated directly with *Bcl-2* expression; the more *Bcl-2* protein the cell line expressed, the more sensitive it was [69]. They also showed that venetoclax sensitivity was not predictable by the *Bcl-2* expression alone, as other pro- and anti-apoptotic proteins also contributed to the sensitivity, as well as the expression of different isoforms, PTM, and subcellular localization. They and others, showed that *Mcl-1*, another pro-survival protein, has been linked to venetoclax resistance and that suppression of *Mcl-1* has been implicated as the mechanistic basis of the synergy [70]. This illustrates, that if we could identify a panel of proteins and PTM, that predict venetoclax sensitivity and if we are able to measure its expression at time of diagnosis, this could rationally select patients that would have a high *a priori* chance to benefit from venetoclax, as well as patients that may need additional inhibition of *Mcl-1*. Figure 3 shows relative protein expression of nine proteins involved in the *BH3*-apoptosis pathway compared to $CD34^+$ cells from healthy donors, and supports the concept that different protein expression patterns exist and that venetoclax response might correlate with one of them.

The same may hold true for the Lysine-specific demethylase 1 (*LSD1*) inhibitors and the Bromodomain and extra-terminal (*BET*) inhibitors that specifically target proteins that play a role in the epigenetic regulation of gene transcription. Both proteins were differentially expressed among patients in our 'histone modification' PFG analysis and patients with high expression of both proteins, individually or in combination, conferred poor prognosis. If we thus can identify which patients express higher levels of *LSD1* or *BRD4* at time of diagnosis, we likely could select patients that are more sensitive to those target therapies which can greatly contribute to rational selection of the right patients for the right treatments.

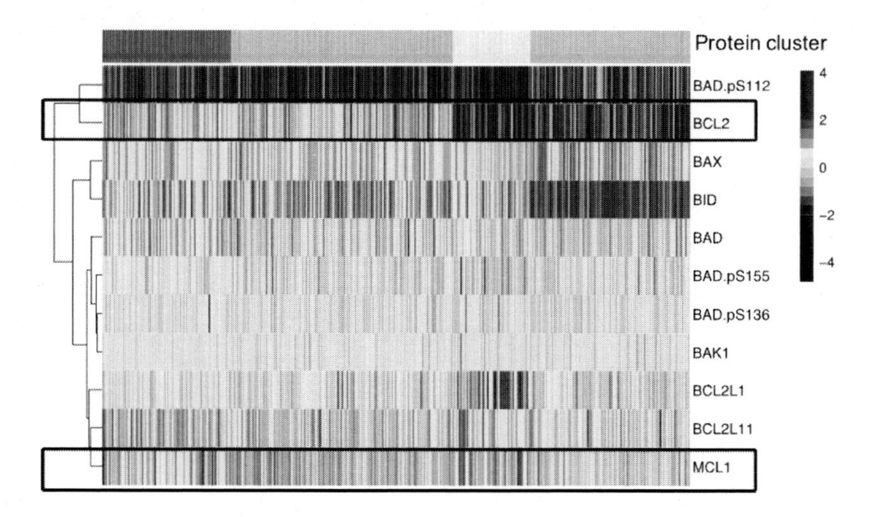

Figure 3. '*Apoptosis-BH3*' protein functional group analysis

Relative protein expression in adult acute leukemia samples for the 9 protein members of the *BH3*-apoptosis protein functional group, including *Bcl-2* and *Mcl-1*. Progeny clustering algorithm determined an optimal number of 4 protein clusters delineated by the 4 colors along the top bar. This heatmap shows variable expression of *Bcl-2*, *Mcl-1* and other *BH3*-members, supporting the concept that different expression patterns exist of which one of them might correlate with venetoclax response.

5.3. Future Perspective; What Else Should We Study?

Although having great promise, certain studies should help optimize the use of proteomics in the future. The first question that should be considered is the use of an appropriate control cell to define the range of 'normal' expression. While the most frequently used control sample is the CD34$^+$ cell population, it would even be more preferable to use normal bone marrow CD34$^+$ CD38$^-$ stem cells to compare against the AML blast. However, as only a very minor fraction of the mononuclear cells in the bone marrow is a hematopoietic stem cell, this complicates its use. Secondly, it would be interesting to study the protein expression in blast AML cells derived from therapy resistant patients or from relapsed AML patients, and compare those to cell from patients that were relapse-free. Evidence suggests that despite the ability of chemotherapy to kill the vast majority of leukemic cells, the rare leukemic stem cell that survives the chemotherapy, is responsible for the outgrowth of the leukemia cells which is manifested as relapse or primary resistant disease. Proteomic analysis of these cells might be more informative than the analysis of the bulk leukemia population, but without a current means to *a priori* identify those few cells, isolation of (enough of) those cells remains a real challenge.

Another interesting approach would be to study proteomics in single cells, rather than studying the average abundance of protein expression across bulk AML cells. Although it was expected that the same genome and the same environment give rise to identical proteomes, it is known that individual cells differ in their proteomes, with sometimes significantly different functional consequences. As described by Simpson's paradox, it is likely true that if you take a population-average protein level, there seems to be a particular correlation between proteins, whereas if you look at the individual cells, proteins can have the same correlation in one cell but the inverse relationship in the other [71] So, if we can detect and study such differences between cells from a single patient, this can provide useful information about cell to cell heterogeneity. Mass cytometry time-of-flight (CyTOF) has already enabled to characterize rare AML stem/progenitor cells in AML and to identify proteins that play key role in drug resistance and AML relapse [72]. Recently Palii et al. utilized this technique to capture temporal dynamics of protein expression of transcription factors in individual cells during human erythropoiesis, showing that most transcription factors from different lineages are co-expressed and that their expression changes gradually during differentiation rather than abruptly [73]. However, so far this method is not cost-effective to perform on large scale. A variant of this would be to study single cell proteomics in different cell populations, for minimal residual disease detection or in cells before and after chemotherapy exposure to evaluate the effect of treatment on the protein expression or activation dynamic over time. Knowing how different cells respond to chemotherapy, would then likely to raise new biological questions about why different cells behave differently, and why or how cells are able to circumvent chemotherapy.

5.4. End Note

We are experiencing an exponential rise in big data generation and dataset availability, however integration of datasets, both of the same type (e.g., protein with protein, mRNA with mRNA), is just beginning to be performed and integration across material type is even more complicated. Complicating integration are differences in methodology between laboratories and platforms. The most used and reviewed 'omics' approaches are the genetic code along with mRNA expression data which, in AML, is indisputably helpful in risk stratification but unfortunately less significant in target therapy selection [74.] As most developed drugs target proteins rather than genes, studying the proteome at time of diagnoses as well as post-treatment hold promise to eventually support the realization of precision medicine. The integration of datasets at larger scales is necessary to achieve such goals, but we believe that there is great potential in combining (epi)genomic, transcriptomic and proteomic data, especially in heterogeneous diseases such as leukemia.

REFERENCES

[1] Döhner H., Weisdorf D. J., Bloomfield C. D. 2015. "Acute Myeloid Leukemia." *N Engl J Med* 373 (12): 1136-1152.

[2] Zhong J. F., Zhao Y., Sutton S., Su A., Zhan Y., Zhu L., Yan C., et al. 2005. "Gene Expression Profile of Murine Long-Term Reconstituting vs. Short-Term Reconstituting Hematopoietic Stem Cells." *Proc Natl Acad Sci USA* 102 (7): 2448-2453.

[3] Ivanova N. B., Dimos J. T., Schaniel C., Hackney J. A., Moore K. A, Lemischka I. R. 2002. "A Stem Cell Molecular Signature." *Science* 298 (5593): 601-604.

[4] Unwin R. D., Smith D. L., Blinco D., Wilson C. L., Miller C. J., Evans C. A., Jaworska E., et al. 2006. "Quantitative Proteomics Reveals Posttranslational Control as a Regulatory Factor in Primary Hematopoietic Stem Cells." *Blood* 107 (12): 4687-4694.

[5] Jassinskaja M., Johansson E., Kristiansen T. A., Åkerstrand H., Sjöholm K., Hauri S., Malmström J., Yuan J., Hansson J. 2017. "Comprehensive Proteomic Characterization of Ontogenic Changes in Hematopoietic Stem and Progenitor Cells." *Cell Rep* 21 (11): 3285-3297.

[6] Ley T. J., Miller C., Ding L., Raphael B. J., Mungall A. J., Robertson A. G., Hoadley K., et al. 2013. "Genomic and Epigenomic Landscapes of Adult De Novo Acute Myeloid Leukemia." *N Engl J Med* 368 (22): 2059-2074.

[7] Papaemmanuil E., Gerstung M., Bullinger L., Gaidzik V. I., Paschka P., Roberts N. D., Potter N. E., et al. 2016. "Genomic Classification and Prognosis in Acute Myeloid Leukemia." *N Engl J Med* 374 (23): 2209-2221.

[8] de Rooij J. D. E., Zwaan C. M., van den Heuvel-Eibrink M. 2015. "Pediatric AML: From Biology to Clinical Management." *J Clin Med* 4 (1): 127-149.

[9] Arber D. A., Orazi A., Hasserjian R., Thiele J., Borowitz M. J., Le Beau M. M., Bloomfield C. D., Cazzola M., Vardiman J. W. 2016. "The 2016 Revision to the World Health Organization Classification

of Myeloid Neoplasms and Acute Leukemia." *Blood* 127 (20): 2391-2405.

[10] Döhner H., Estey E. H., Amadori S., Appelbaum F. R., Büchner T., Burnett A. K., Dombret H., et al. 2010. "Diagnosis and Management of Acute Myeloid Leukemia in Adults: Recommendations from an International Expert Panel, on Behalf of the European LeukemiaNet." *Blood* 115 (3): 453-474.

[11] Grimwade D., Walker H., Oliver F., Wheatley K., Harrison C., Harrison G., Rees J., et al. 1998. "The Importance of Diagnostic Cytogenetics on Outcome in AML: Analysis of 1,612 Patients Entered into the MRC AML 10 Trial the Medical Research Council Adult and Children's Leukaemia Working Parties." *Blood* 92 (7): 2322.

[12] Verhaak R., Goudswaard C., Putten W., Bijl M., Sanders M., Hugens W., Uitterlinden A., et al. 2005. "Mutations in Nucleophosmin (NPM1) in Acute Myeloid Leukemia (AML): Association with Other Gene Abnormalities and Previously Established Gene Expression Signatures and their Favorable Prognostic Significance." *Blood* 106 (12): 3747-3754.

[13] Figueroa M. E., Omar A. W., Lu C., Ward P. S, Patel J., Shih A., Li Y., et al. 2010. "Leukemic IDH1 and IDH2 Mutations Result in a Hypermethylation Phenotype, Disrupt TET2 Function, and Impair Hematopoietic Differentiation." *Cancer Cell* 18 (6): 553-567.

[14] Stone R. M., Mandrekar S. J., Sanford B. L., Laumann K., Geyer S., Bloomfield C. D., Thiede C., et al. 2017. "Midostaurin Plus Chemotherapy for Acute Myeloid Leukemia with a FLT3 Mutation." *N Engl J Med* 377 (5): 454-464.

[15] Levis M. 2017. "Midostaurin Approved for FLT3-Mutated AML." *Blood* 129 (26): 3403-3406.

[16] Estey E. 2015. "Why is Progress in Acute Myeloid Leukemia so Slow?" *Semin Hematol* 52 (3): 243-248.

[17] Genshaft A. S., Li S., Gallant C. J., Darmanis S., Prakadan S. M., Ziegler C. G. K., Lundberg M., et al. 2016. "Multiplexed, Targeted Profiling of Single-Cell Proteomes and Transcriptomes in a Single Reaction." *Genome Biol* 17 (1): 188.

[18] Vogel C., Marcotte E. M. 2012. "Insights into the Regulation of Protein Abundance from Proteomic and Transcriptomic Analyses." *Nat Rev Genet* 13 (4): 227-232.

[19] Payne S. H. 2015. "The Utility of Protein and mRNA Correlation." *Trends in Biochem Sci* 40 (1): 1-3.

[20] Roumeliotis T. I., Williams S. P., Gonçalves E., Alsinet C., Del Castillo Velasco-Herrera M., Aben N., Ghavidel F. Z., et al. 2017. "Genomic Determinants of Protein Abundance Variation in Colorectal Cancer Cells." *Cell Rep* 20 (9): 2201-2214.

[21] Zhang B., Wang J., Wang X., Zhu J., Liu Q., Shi Z., Chambers M. C., et al. 2014. "Proteogenomic Characterization of Human Colon and Rectal Cancer." *Nature* 513 (7518): 382-387.

[22] Mun D. G., Bhin J., Kim J., Kim S., Kim H. K., Kim D. H., Kim K. P., et al. 2019. "Proteogenomic Characterization of Human Early-Onset Gastric Cancer." *Cancer Cell* 35 (1): 124.e10.

[23] Mertins P., Mani D. R., Ruggles K. V., Gillette M. A., Clauser K. R., Wang P., Wang X., et al. 2016. "Proteogenomics Connects Somatic Mutations to Signalling in Breast Cancer." *Nature* 534 (7605): 55-62.

[24] Archer T. C., Ehrenberger T., Mundt F., Gold M. P., Krug K., Mah C. K., Mahoney E. L., et al. 2018. "Proteomics, Post-Translational Modifications, and Integrative Analyses Reveal Molecular Heterogeneity within Medulloblastoma Subgroups." *Cancer Cell* 34 (3): 410.e8.

[25] Harper J. W., Bennett E. J. 2016. "Proteome Complexity and the Forces that Drive Proteome Imbalance." *Nature* 537 (7620): 328-338.

[26] Aebersold R., Mann M. 2003. "Mass Spectrometry-Based Proteomics." *Nature* 422 (6928): 198-207.

[27] Hillenkamp F., Karas M., Beavis R. C., Chait B. T. 1991. "Matrix-Assisted Laser Desorption/Ionization Mass Spectrometry of Biopolymers." *Anal Chem* 63 (24): 1203A.

[28] Tanaka K., Waki H., Ido Y., Akita S., Yoshida Y., Yoshida T., Matsuo T. 1988. "Protein and Polymer Analyses Up Tom/Z 100 000 by Laser Ionization Time-of-Flight Mass Spectrometry." *Rapid Commun Mass Spectrom* 2 (8): 151-153.

[29] Issaq H. J., Veenstra T. D., Conrads T. P., Felschow D. 2002. "The SELDI-TOF MS Approach to Proteomics: Protein Profiling and Biomarker Identification." *Biochem Biophys Res Commun* 292 (3): 587-592.

[30] Whitehouse C. M., Dreyer R. N., Yamashita M., Fenn J. B. 1989. "Electrospray Ionization for Mass Spectrometry." *Science* 246 (4926): 64-71.

[31] Nicolas E., Ramus C., Berthier S., Arlotto M., Bouamrani A., Lefebvre C., Morel F., et al. 2011. "Expression of S100A8 in Leukemic Cells Predicts Poor Survival in De Novo AML Patients." *Leukemia* 25 (1): 57-65.

[32] Xu Y., Zhuo J., Duan Y., Shi B., Chen X., Zhang X., Xiao L., et al. 2014. "Construction of Protein Profile Classification Model and Screening of Proteomic Signature of Acute Leukemia." *Int J Clin Exp Pathol* 7 (9): 5569.

[33] Block H., Maertens B., Spriestersbach A., Brinker N., Kubicek J., Fabis R., Labahn J., Schäfer F. 2009. "Immobilized-Metal Affinity Chromatography (IMAC): A Review." *Methods in Enzymol* 463: 439.

[34] Vidova V., Spacil Z. 2017. "A Review on Mass Spectrometry-Based Quantitative Proteomics: Targeted and Data Independent Acquisition." *Anal Chim Acta* 964: 7-23.

[35] Matondo M., Marcellin M., Chaoui K., Bousquet-Dubouch M. P., Gonzalez-de-Peredo A., Monsarrat B., Burlet-Schiltz O. 2017. "Determination of Differentially Regulated Proteins upon Proteasome Inhibition in AML Cell Lines by the Combination of Large-scale and Targeted Quantitative Proteomics." *Proteomics* 17 (7)

[36] Zhu W., Smith J. W., Huang C. M. 2010. "Mass Spectrometry-Based Label-Free Quantitative Proteomics." *J Biomed Biotechnol* 2010: 840518-6.

[37] Zhang G., Neubert T. A. 2009. "Use of Stable Isotope Labeling by Amino Acids in Cell Culture (SILAC) for Phosphotyrosine Protein Identification and Quantitation." *Methods Mol Biol* 527: 79.

[38] Ong S. E., Blagoev B., Kratchmarova I, Kristensen D. B., Steen H., Pandey A., Mann M. 2002. "Stable Isotope Labeling by Amino Acids

in Cell Culture, SILAC, as a Simple and Accurate Approach to Expression Proteomics." *Mol Cell Proteomics* 1 (5): 376-386.

[39] Sandow J. J., Infusini G., Holik A. Z., Brumatti G., Averink T. V., Ekert P. G., Webb A. I. 2017. "Quantitative Proteomic Analysis of EZH2 Inhibition in Acute Myeloid Leukemia Reveals the Targets and Pathways that Precede the Induction of Cell Death." *Proteomics Clin Appl* 11 (9-10).

[40] Galli S., Zlobec I., Schürch C., Perren A., Ochsenbein A. F., Banz Y. 2015. "CD47 Protein Expression in Acute Myeloid Leukemia: A Tissue Microarray-Based Analysis." *Leuk Res* 39 (7): 749-756.

[41] Mattsson K., Honkaniemi E., Barbany G., Gustafsson B. 2015. "Increased p53 Protein Expression as a Potential Predictor of Early Relapse After Hematopoietic Stem Cell Transplantation in Children with Acute Myelogenous Leukemia." *Pediatr Transplantat* 19 (7): 767-775.

[42] Zhou J., Chng W. J. 2014. "Identification and Targeting Leukemia Stem Cells: The Path to the Cure for Acute Myeloid Leukemia." *World J Stem Cells*. 6 (4): 473-484.

[43] Pollyea D. A., Jordan C. T. 2017. "Therapeutic Targeting of Acute Myeloid Leukemia Stem Cells." *Blood* 129 (12): 1627-1635.

[44] Kornblau S. M., Qutub A. A., Yao H., York H., Qiu Y., Graber D., Ravandi F., et al. 2013. "Proteomic Profiling Identifies Distinct Protein Patterns in Acute Myelogenous Leukemia CD34+CD38- Stem-Like Cells." *PLoS One* 8 (10): e78453.

[45] Hu C. W., Qiu Y, Ligeralde A., Raybon A. Y., Yoo S. Y., Coombes K. R., Qutub A. A., Kornblau S. M. 2019. "A Quantitative Analysis of Heterogeneities and Hallmarks in Acute Myelogenous Leukaemia." *Nat Biomed Eng*.

[46] Braoudaki M., Tzortzatou-Stathopoulou F., Anagnostopoulos A., Papathanassiou C., Vougas K., Karamolegou K., Tsangaris G. 2011. "Proteomic Analysis of Childhood De Novo Acute Myeloid Leukemia and Myelodysplastic Syndrome/AML: Correlation to Molecular and Cytogenetic Analyses." *Amino Acids* 40 (3): 943-951.

[47] Kornblau S. M., Coombes K. R. 2011. "Use of Reverse Phase Protein Microarrays to Study Protein Expression in Leukemia: Technical and Methodological Lessons Learned." *Methods Mol Biol* 785: 141.

[48] Butler J. S., Qiu Y., Zhang N., Yoo S. Y., Coombes K. R., Dent S. Y. R., Kornblau S. M. 2017. "Low Expression of ASH2L Protein Correlates with a Favorable Outcome in Acute Myeloid Leukemia." *Leuk Lymphoma* 58 (5): 1207-1218.

[49] Kornblau S. M., Qiu Y., Zhang N., Singh N., Faderl S., Ferrajoli A., York H., Qutub A. A., Coombes K. R., Watson D. K. 2011. "Abnormal Expression of FLI1 Protein is an Adverse Prognostic Factor in Acute Myeloid Leukemia." *Blood* 118 (20): 5604-5612.

[50] Kornblau S. M., Singh N., Qiu Y., Chen W., Zhang N., Coombes K. R. 2010. "Highly Phosphorylated FOXO3A is an Adverse Prognostic Factor in Acute Myeloid Leukemia." *Clin Canc Res* 16 (6): 1865-1874.

[51] Ruvolo V. R., Ruvolo P. P., Burks J. K., Qiu Y., Wang R., Shpall E. J., Mirandola L., et al. 2018. "Role of MSC-Derived Galectin 3 in the AML Microenvironment." *Biochim Biophys Acta Mol Cell Res* 1865 (7): 959-969.

[52] Pierce A., Whetton A. D., Meyer S., Ravandi F., Borthakur G, Coombes K. R., Zhang N., Kornblau S. M. 2013. "Transglutaminase 2 Expression in Acute Myeloid Leukemia: Association with Adhesion Molecule Expression and Leukemic Blast Motility." *Proteomics* 13 (14): 2216-2224.

[53] Quintás-Cardama A., Zhang N., Qiu Y., Post S. M., Creighton C., Cortes J., Coombes K. R., Kornblau S. M. 2015. "Loss of TRIM62 Expression is an Independent Adverse Prognostic Factor in Acute Myeloid Leukemia." *Clin Lymphoma Myeloma* 15 (2): 127.e15.

[54] Kornblau S. M., Tibes R., Qiu Y., Chen W., Kantarjian H. M., Andreeff M., Coombes K. R., Mills G. B. 2009. "Functional Proteomic Profiling of AML Predicts Response and Survival." *Blood* 113 (1): 154-164.

[55] Kornblau S. M., Ruvolo P. P., Wang R. Y., Battula V. L., Shpall E. J., Ruvolo V. R., McQueen T., et al. 2018. "Distinct Protein Signatures

of Acute Myeloid Leukemia Bone Marrow-Derived Stromal Cells are Prognostic for Patient Survival." *Haematologica* 103 (5): 810-821.

[56] Quintás-cardama A., Hu C. W., Qutub A. A., Qiu Y., Zhang X., Post S. M., Zhang N., Coombes K. R., Kornblau S. M. 2017. "p53 Pathway Dysfunction is Highly Prevalent in Acute Myeloid Leukemia Independent of TP53 Mutational Status." *Leukemia* 31 (6): 1296-1305.

[57] van Dijk A. D., Hu C. W., de Bont, E. S. J. M, Qiu Y., Hoff F. W., Yoo S. Y Coombes., K. R., Qutub A. A., Kornblau S. M. 2018. "Histone Modification Patterns using RPPA-Based Profiling Predict Outcome in Acute Myeloid Leukemia Patients." *Proteomics* 18 (8).

[58] Hu C. W., Kornblau S. M., Slater J. H., Qutub A. A. 2015. "Progeny Clustering: A Method to Identify Biological Phenotypes." *Sci Rep* 5 (1): 12894.

[59] Barth T. K., Imhof A. 2010. "Fast Signals and Slow Marks: The Dynamics of Histone Modifications." *Trends Biochem Sci* 35 (11): 618-626.

[60] Valencia A, M., Kadoch C. 2019. "Chromatin Regulatory Mechanisms and Therapeutic Opportunities in Cancer." *Nat Cell Biol* 21 (2): 152-161.

[61] Lafon-Hughes L., Di Tomaso M. V., Méndez-Acuña L., Martínez-López W. 2008. "Chromatin-Remodelling Mechanisms in Cancer." *Mutat Res* 658 (3): 191.

[62] Fiziev P., Akdemir K. C., Miller J. P., Keung E. Z., Samant N. S., Sharma S., Natale C. A., et al. 2017. "Systematic Epigenomic Analysis Reveals Chromatin States Associated with Melanoma Progression." *Cell Reports* 19 (4): 875-889.

[63] Cedar H., Bergman Y. 2009. "Linking DNA Methylation and Histone Modification: Patterns and Paradigms." *Nat Rev Genetics* 10 (5): 295-304.

[64] Govaert G., Nadif M. 2003. "Clustering with Block Mixture Models." *Pattern Recognition* 36 (2): 463-473.

[65] Hu C. W., Qiu Y., Ligeralde A., Raybon A. Y., Yoo S. Y., Coombes K. R., Qutub A. A., Kornblau S. M. 2019. "A Quantitative Analysis of

Heterogeneities and Hallmarks in Acute Myelogenous Leukaemia." *Nat Biom Eng*.

[66] Hoff F. W., Hu C. W., Qutub A. A., Qiu Y., Hornbaker M. J., Bueso-Ramos C., Abbas H. A., Post S. M., de Bont, E. S. J. M., Kornblau S. M. 2019. "Proteomic Profiling of Acute Promyelocytic Leukemia Identifies Two Protein Signatures Associated with Relapse." *Proteomics Clin Appl*: e1800133.

[67] Hoff F. W., Hu. C. W., Qiu Y., Ligeralde A., Yoo S. Y., Mahmud M., de Bont, E. S. J. M., Qutub A. A., Horton T. M., Kornblau S. M. 2018. "Recognition of Recurrent Protein Expression Patterns in Pediatric Acute Myeloid Leukemia Identified New Therapeutic Targets." *Mol Canc Res* 16 (8): 1275-1286.

[68] Hoff F. W., Hu C. W., Qiu Y., Ligeralde A., Yoo S. Y., Scheurer M. E., de Bont, E. S. J. M., Qutub A. A., Kornblau S. M., Horton T. M. 2018. "Recurrent Patterns of Protein Expression Signatures in Pediatric Acute Lymphoblastic Leukemia: Recognition and Therapeutic Guidance." *Mol Canc Res* 16 (8): 1263-1274.

[69] Souers A. J., Leverson J. D., Boghaert E. R., Ackler S. L., Catron N. D., Chen J., Dayton B. D., et al. 2013. "ABT-199, a Potent and Selective BCL-2 Inhibitor, Achieves Antitumor Activity while Sparing Platelets." *Nat Med* 19 (2): 202-208.

[70] Teh T-C, Nguyen N-Y, Moujalled D. M., Segal D., Pomilio G., Rijal S., Jabbour A., et al. 2018. "Enhancing Venetoclax Activity in Acute Myeloid Leukemia by Co-Targeting MCL1." *Leukemia* 32 (2): 303-312.

[71] Blyth, C. R. 1972. "On Simpson's Paradox and the Sure-Thing Principle." *J Am Stat Assoc* 67 (338): 364-366.

[72] Zeng Z., Konopleva M., Andreeff M. 2017. "Single-Cell Mass Cytometry of Acute Myeloid Leukemia and Leukemia Stem/Progenitor Cells." *Methods Mol Biol* 1633: 75.

[73] Palii C. G., Cheng Q., Gillespie M. A., Shannon P., Mazurczyk M., Napolitani G., Price N. D., et al. 2019. "Single-Cell Proteomics Reveal that Quantitative Changes in Co-Expressed Lineage-Specific

Transcription Factors Determine Cell Fate." *Cell Stem Cell* 24 (5): 820.e5.

[74] López de Maturana, E., Alonso L., Alarcón P., Martín-Antoniano I. A., Pineda S., Piorno L., Luz Calle M., Malats N. 2019. "Challenges in the Integration of Omics and Non-Omics Data." *Genes* 10 (3): 238.

In: Proteomics
Editor: Ricardo Parker

ISBN: 978-1-53616-440-4

Chapter 2

MOLECULAR MECHANISM OF OSELTAMIVIR AND OTHER LEAD COMPOUNDS TOWARDS INFLUENZA A VIRUS BASED ON PROTEOMICS AND TRANSCRIPTOMICS APPROACHES

Usman Sumo Friend Tambunan[1,*]
and Arli Aditya Parikesit[2]
[1]Bioinformatics Research Group, Department of Chemistry, Faculty of Mathematics and Sciences, University of Indonesia, Depok, Indonesia
[2]Department of Bioinformatics, School of Life Sciences, Indonesia International Institute for Life Sciences, East Jakarta, Indonesia

ABSTRACT

The influenza A virus is one of the most dangerous pathogens threatening human life. One of the default treatments for influenza A virus

* Corresponding Author's E-mail: usman@ui.ac.id.

is prescribing the drug oseltamivir. However, due to the mutation rate of influenza A as an RNA-based virus, there are tendencies of resistance toward oseltamivir. In this respect, novel approaches should be devised to ward off the menace of influenza A, mainly with modern proteomics- and transcriptomics-based methods. This review will first discuss the molecular mechanism of oseltamivir and the advantages and disadvantages of using it. Proteomics- and transcriptomics-based alternatives to oseltamivir will be described as well. Our proteomics-based computational approach has successfully produced some fine-grained designs, such as cyclic peptide-based, flavonoid-based, and amantadine-derivative leads. The focus going forward should be on the promising design of cyclic peptide because of its stability in our physiological system. At the same time, the availability of natural products based on the biodiversity database and the derivative products of the existing drug should not be overlooked. However, another approach is being developed as well, based upon the transcriptomics method, where, the coding and non-coding RNAs are utilized as drug candidates. The most common drug candidates in use are silencing (si)RNA, which is a short-strand, non-coding RNA. The issues plaguing RNA-based drugs are mainly the delivery mode to the target, as this molecule tends to be unstable in the cell. In this regard, the best approach always depends on the diagnostic results drawn by medical doctors. Despite all of this, oseltamivir remains the default medication for influenza A, in combination with other drugs such as zanamivir.

Keywords: oseltamivir, transcriptomics, proteomics, non-coding RNA, influenza

INTRODUCTION

Influenza A-based disease is considered one of the most hazardous infectious diseases in the world (WHO 2014, 2012). It also greatly affects several countries in southeast Asia, with Indonesia being the most affected region with the highest mortality rate (Kemenkes-RI 2013). This review focuses particular attention on the influenza A virus over other strains, such as influenza B and C, because it has the most diverse variability due to its molecular evolution (Davis 2014). In this regard, the world has already seen the massive mortality caused by Spanish influenza with millions of deaths in 1918 (Taubenberger and Morens 2006). The well-known vectors for influenza A virus are swine, bat, and avian (birds and poultry) (Behrens et al. 2006). The influenza A virus is contagious via animal-to-person transmission (WHO 2012). Hence, some very limited person-to-person transmissions were also observable by epidemiologist and laboratory reports, but no further transmission to third persons or parties is feasible (Gao et al. 2013; Ungchusak et al. 2005; Kandun et al. 2006).

This virus has also become pandemic in the sense that it is spread worldwide and most people don't have sufficient immunity against it (WHO 2015). Although due to the advancement of medical technology, the mortality of the influenza A virus is greatly reduced, it is still considered highly hazardous to the human population. Scientists are trying to do away with the classical approach of medicine and embrace modern molecular-based medication in order to tackle this menace (Davidov et al. 2003). A breakthrough was reached when the Human Genome Project (HGP) was completed in 2001, which opened the door to sequencing other organisms, such as viruses, for medical research (Butler 2001; Collins and McKusick 2001). Thus, the genome of the influenza A virus was unraveled in an exact manner (Tambunan and Parikesit 2010).

The molecular architecture of the influenza A virus is considered standard, as other RNA-based ones are. The genome architecture of the influenza A virus is comprised of eight different segments, each encoded with its own protein with a length of less than 800 amino acids (Naulet et al. 2018). The length of the influenza A viral genome is 13,588 bases long and it is considered a small-sized genome (Winter and Fields 1982). This viral genomics information was eventually stored in the GenBank/NCBI database along with other organisms (Wheeler et al. 2007). The GenBank database was employed for easy retrieval of genomics, proteomics, and transcriptomics information of the influenza A virus that eventually will be useful for drug design. As the development of GenBank database reached its peak with employing advances in computer sciences such as 'big data' and utilizing new programming language such as Python, new field of study, namely bioinformatics, was born. It is defined as multidisciplinary science between biology and computer science (Welch et al. 2014). In this regard, bioinformatics will play an important role in drug development for influenza A virus infections.

Haemaglutinin is the protein responsible for the attachment of the virus to the host cell, and neuraminidase is the protein responsible for the detachment of the virus from the host cell (Behrens et al. 2006). Hence, the naming convention of the virus is based on its abbreviation, H (haemaglutinin) and N (neuraminidase). Notorious examples of viruses with this naming convention include avian influenza (H5N1), swine influenza (H1N1), and novel avian-origin influenza (H7N9). The H5- and H7-based viruses are classified as highly pathogenic avian influenza viruses (HPAI).

A standard practice in warding off the influenza A virus is to inhibit certain proteins and/or enzymes involved in the viral replication or attachment to the host cell (Day and Cohen 2013; Yagi et al. 2007; Sattar et al. 2004). This method has already proven successful in other viral infections such as the development of the antiretroviral drug for HIV/AIDS (Yanuar et al. 2014; Syahdi et al. 2012). A standard bioassay to test the inhibition efficacy of anti-viral drugs is already developed, as is a computational method to support the assay (Tambunan and Parikesit 2013; Luzhkov et al. 2007).

The inhibition assay method has produced oseltamivir as a standard drug to ward off HPAI (Sugrue et al. 2008). It works by inhibiting the neuraminidase (NA) enzyme in the virus (Sugrue et al. 2008). Earlier, it was included in the WHO's list of essential (core) medicines (WHO 2010). In the early days of molecular medicine research, oseltamivir was considered an indispensable drug in influenza A research.

However, mutations in the HPAI, especially in the NA protein, have reduced the efficacy of oseltamivir (Ferraris and Lina 2008). This happens because HPAI is an RNA-based virus that evolved very quickly due to the instability of its RNA-based genome (Marz et al. 2014). Solid evidence of oseltamivir resistance made the WHO remove it from their 'core' drug list (Kmietowicz 2017). Since then, the WHO and scientific communities have been trying to develop alternative drugs to oseltamivir that could be applied to efficiently tackle the influenza A infection. Bioinformatics is one of the most important fields of study to play a part in developing such alternatives.

In this respect, novel approaches to ward off HPAI are necessary. A proteomics-based approach was developed to provide mechanistic insight into protein interactions in the host cell (Nakamura 2007; Uhlén et al. 2015; X. Wu, Hasan, and Chen 2014), while a transcriptomics approach was developed by providing RNA interaction with the biomarkers (Dong and Chen 2013; Gomase 2009). Thus, a systemic biology-based approach is being developed as well by integrating the fields of genomics, transcriptomics, and proteomics into one holistic approach (Butcher, Berg, and Kunkel 2004).

The objective of this review is to provide information on the molecular basis of oseltamivir and show it is a viable alternative in the scope of proteomics- and transcriptomics-based approaches. It is expected that the discussion induced from this review will eventually enrich the intellectual contribution in influenza A research especially in relation with the application of oseltamivir and its alternatives.

MOLECULAR MECHANISM OF OSELTAMIVIR

Oseltamivir is an acetamido cyclohexene that is a structural homolog of sialic acid and inhibits neuraminidase (Pubchem 2018; Drugbank 2018). Oseltamivir is active by its biotransformation product, the oseltamivir carboxylate (LI, WANG, and CHEN 2013). The ethyl ester prodrug formation of oseltamivir will produce its phosphate form, which eventually will be hydrolyzed into the reactive substance as a neuraminidase inhibitor (Hama 2015). Oseltamivir is known to be active towards such target molecules as the adenosine A receptor, norepinephrine receptor, sigma receptor, and serotonin receptor (Lindemann et al. 2010). This phenomenon should be observed with care as alteration to normal cell physiology is feasible (Muraki et al. 2015; Hiasa et al. 2013).

From the perspective of computational organic chemistry, the possible mechanism for ester and amide hydrolysis of oseltamivir is with a stepwise mechanism that favors tetrahedral intermediate formation (Li, Wang, and Chen 2013). It is also known that neuramindase is an exosialidase enzyme

(EC 3.2.1.18) that could be inhibited by oseltamivir (Shtyrya, Mochalova, and Bovin 2009). In this respect, the standard lock-and-key of enzymatic mechanism is applied. Moreover, further study of the molecular dynamics of mutation in the neuramindase enzyme exposees a weakening of the hydrogen bond, although no significant change in conformation has been observed. However, a weakening sensitivity to oseltamivir has been observed (Takano et al. 2013).

The interesting molecular mechanism of oseltamivir occurred during its hydrolysis reaction. In terms of toxicity, the hydrolytic product of oseltamivir will have increased impact. However, prevention of increased toxicity could be guaranteed with the hydrolysis inhibition (Shi et al. 2006).

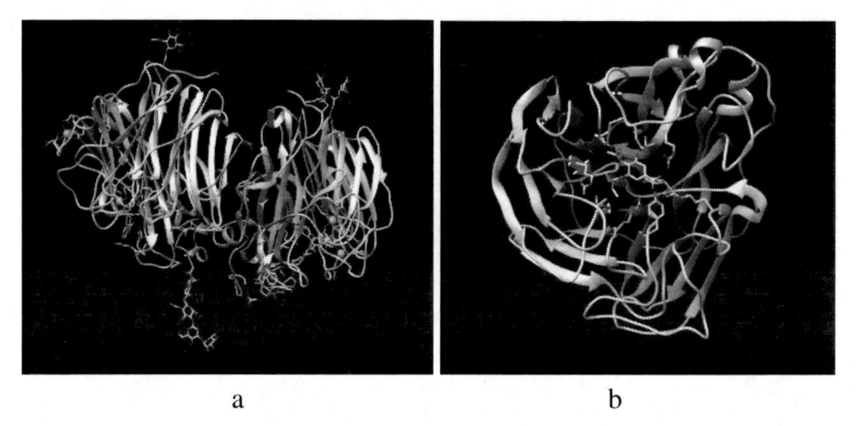

a b

Source: http://www.rcsb.org/structure.

Figure 1. A) Crystal structure of 2009 pandemic H1N1 neuraminidase complexed with oseltamivir (PDB ID: 3TI6) B) N8 neuraminidase in complex with oseltamivir (PDB ID: 2HT8). The molecular visualization was provided with CHIMERA software version release 1.13.1.

Based upon recent bioinformatics research, it was found that the neuraminidase enzyme form was clustered into two distinct clusters, namely group-1 and group-2 (Russell et al. 2006; Vavricka et al. 2011). Group-1 comprises N1, N4, and N8. The N8 protein interaction with oseltamivir is shown in Figures 1 a and b, as the oseltamivir drug is attached to the 150-loop region of the protein. The lead compound-protein interaction in this

150-loop region is the starting point for the future development of an influenza A anti-viral drug.

The underlying cavity of the neuraminidase protein is the target for the lead compounds. However, the parameters that should be observed are not only the existence of the cavity or crevice, but also the physico-chemical parameters of the protein-lead compound interaction (Jónsdóttir, Jørgensen, and Brunak 2005). It mainly emphasizes the chemical bonds among them, with the most notable examples being the covalent bond, hydrogen bond, and Van der Waals interaction (MacKerell et al. 1998). Moreover, the toxicity and pharmacology-based parameters are considered as well, such as the ADME-TOX (absorption, distribution, metabolism, and toxicity) parameters (Shaikh et al. 2007). In this regard, the oseltamivir drug already passed the requirements of the design parameters and has been deemed the most acceptable drug against the influenza A virus (Grienke et al. 2012).

Proteomics- and Transcriptomics-Based Drug Alternatives to Oseltamivir

The mutation of the N9 protein, the R294K, conferred a strong resistance to oseltamivir, but only a mild one toward the drug laninamivir, as shown in Figure 2 b (Y. Wu et al. 2013). Hence, another drug in use for combatting the influenza A virus is zanamivir, as shown in Figure 2 a (Russell et al. 2006; Vavricka et al. 2011). Moreover, natural products for inhibiting neuraminidase were applied in a computational study (A. A. Parikesit et al. 2016) as were oseltamivir derivatives (Kocik et al. 2014; Hsu et al. 2018). The natural product database HERBADB was utilized for the computational study (Yanuar et al. 2011). Hence, the oseltamivir and its derivatives were taken from the PubChem database and curated with ChemSketch software for structural design (Wheeler et al. 2007; Spessard 1998). The existing bioinformatics research shows that alternative drug such as laninamivir and zanamivir definitely perform better than oseltamivir in the virtual screening assay. This situation creates hope among drug designers as new leads are developed. However, caution should be taken as this pipeline would need special expertise in biochemistry, and organic chemistry especially. This is because the main problem in drug development usually relates to the inability of lead compounds to reach the target cell or protein, the bad resolution of the chemical bonding visualization, and/or the inability to visualize the trajectory of the chemical reaction (van Gunsteren et al. 2006; Kumari et al. 2017). The application of biochemistry and organic chemistry in the field of bioinformatics is called 'structural bioinformatics' (Chandra, Anand, and Yeturu 2010).

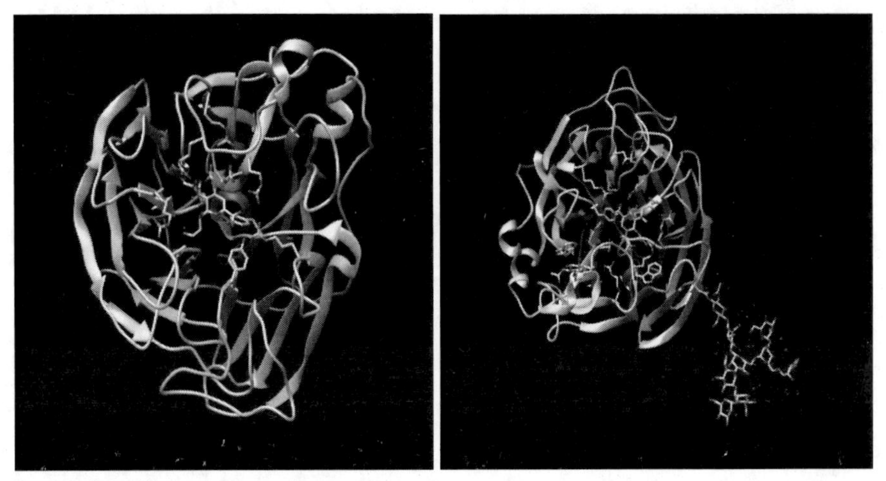

Source: http://www.rcsb.org/structure.

Figure 2. A) N8 neuraminidase in complex with zanamivir (PDB ID: 2HTQ) B) Anhui N9-laninamivir (PDB ID:4MWU). The molecular visualization was provided with Chimera software version release 1.13.1.

Thus, the experiments on oseltamivir and its alternatives were stored in centralized databases such as the DrugBank database (Wishart et al. 2008), which was mapped to the virus annotation database, ViPR, to provide functional features of drug properties. The database provides the information on the utilization of influenza A drugs, and their mechanism of action (pharmacology) is shown in Figure 3 (Pickett et al. 2012).

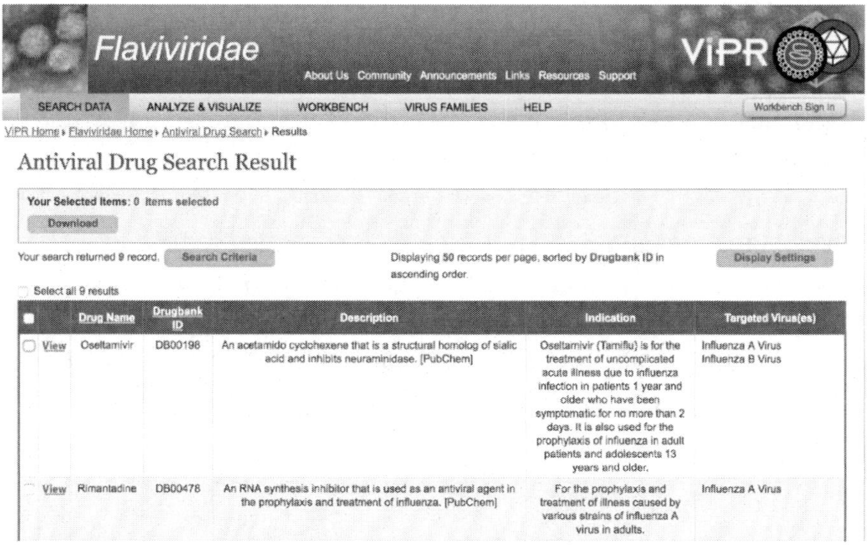

		Drug Name	Drugbank ID	Description	Indication	Targeted Virus(es)
	View	Oseltamivir	DB00198	An acetamido cyclohexene that is a structural homolog of sialic acid and inhibits neuraminidase. [PubChem]	Oseltamivir (Tamiflu) is for the treatment of uncomplicated acute illness due to influenza infection in patients 1 year and older who have been symptomatic for no more than 2 days. It is also used for the prophylaxis of influenza in adult patients and adolescents 13 years and older.	Influenza A Virus Influenza B Virus
	View	Rimantadine	DB00478	An RNA synthesis inhibitor that is used as an antiviral agent in the prophylaxis and treatment of influenza. [PubChem]	For the prophylaxis and treatment of illness caused by various strains of influenza A virus in adults.	Influenza A Virus

Source: https://www.viprbrc.org/.

Figure 3. The search results for antiviral drugs for influenza A in the ViPR (Virus Pathogen Resource) database. It yielded nine drugs, namely oseltamivir, rimantadine, zanamivir, ribavirin, amantadine, peramivir, laninamivir, moroxydine, and umifenovir.

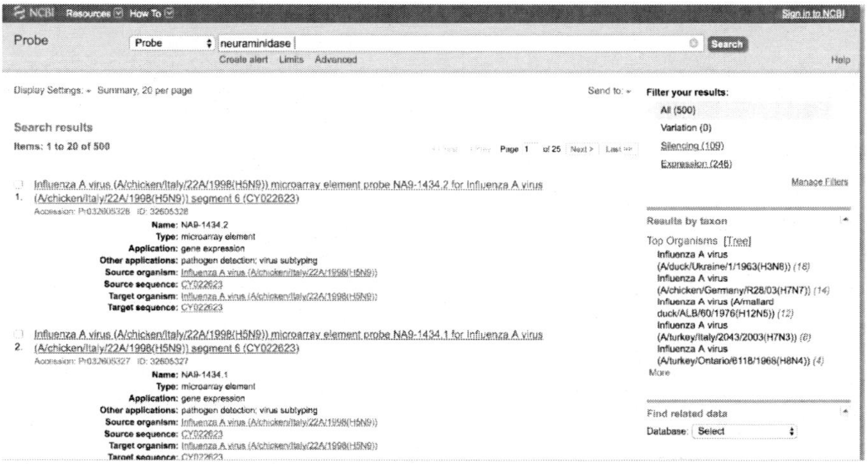

Source: https://www.ncbi.nlm.nih.gov/projects/genome/rnai/.

Figure 4. The silencing (si)RNA database of NCBI. Here the search results of the neuraminidase gene silencer are shown, that is, 500 probe hits.

The ViPR database provides complete antiviral drug information for influenza A by providing the drug name, DrugBank ID, mechanism description, indication, and target virus. However, as the conventional proteomics-based approach seemed to be at a standstill, a novel approach for drug design has risen in the form of transcriptomics studies. The popular method in transcriptomics is developing short-strand non-coding RNA as a gene silencer, named silencing (si)RNA (Hannon 2002). This approach was already applied in cancer drug development and has been a popular method in the area of infectious diseases (Arli Aditya Parikesit 2018; Arli Aditya Parikesit, Utomo, and Karimah 2018). Herewith, shown in Figure 4, NCBI has provided a database for siRNA that works by inhibiting neuraminidase genes (Dykxhoorn and Lieberman 2005; Sætrom and Snøve 2004).

OUR CYCLIC-PEPTIDE-BASED DRUG DEVELOPMENT

Our structural bioinformatics laboratory has successfully designed several cyclic-peptide-based drugs for treating the influenza A virus. We chose to develop peptide-based drugs due to the stability, bioabsorption, and reliability of peptides (Edwards et al. 2007; Shaji and Patole 2008). However, linear peptides tend to be unstable due to the degradation of the protease enzyme, and this situation has made linear designs unreliable (Bogdanowich-Knipp et al. 1999). In this respect, scientists have tried one efficient strategy that has proven to be correct: providing the cyclization of the peptide molecules protects them from the proteolytic enzymes (Soltero 2005). Cyclic-peptide has proven its robustness during in vitro and in vivo studies by reaching the molecular receptor target in an efficient manner (Iskandarsyah et al. 2008; Sinaga et al. 2002; On et al. 2014). This motivates us to develop our own approach in designing cyclic-peptide-based drugs with bioinformatics tools and methodology.

Our approach was mainly motivated to decrease the expensive venue of wet laboratory research and improve the utilization of bioinformatics methods to supplement the gap that was left by wet lab research (Arli Aditya Parikesit 2009). It is already confirmed that bioinformatics research could assist the drug development pipeline by slashing significant costs in laboratory investment (Whittaker 2003; Desany and Zhang 2004; Huang et al. 2010). Thus, Table 1 shows our developed cyclic-peptide drug designs from our bioinformatics pipeline.

Table 1. The cyclic-peptide ligands for influenza A virus from our lab

No.	Ligand	Virus Target	Target Protein	References
1.	CKTTC	H1N1	Polymerase A and B1 (PAC-PB1N) H1N1	(Arli Aditya Parikesit et al. 2014)
2.	CKKTC	H1N1	Polymerase A and B1 (PAC-PB1N) in	(Tambunan et al. 2016)
3.	CKTTC	H1N1	Polymerase A and B1 (PAC-PB1N) in	(Tambunan et al. 2016)
4.	CLDRC	H5N1 (HPAI)	Neuraminidase	(Tambunan et al. 2015)
5.	CIWRC	H5N1 (HPAI)	Neuraminidase	(Tambunan et al. 2015)
6.	NNY	H1N1	Neuraminidase	(Tambunan, Amri, and Parikesit 2012)
7.	DNY	H1N1	Neuraminidase	(Tambunan, Amri, and Parikesit 2012)
8.	CRMYPC	H1N1	Neuraminidase	(Tambunan et al. 2014)
9.	CRNFPC	H1N1	Neuraminidase	(Tambunan et al. 2014)

Our cyclic-peptides were designed using three distinct methods in one integrated structural bioinformatics pipeline (Tambunan et al. 2014). First, the proteins were downloaded and the peptides were designed from a reputable database (NCBI). The cyclic-peptide sequences were designed by combining variations of polar and non-polar amino acids attached with a disulfide bond (Tambunan et al. 2015). Second, a molecular docking method was employed to select the most thermodynamically favorable leads. Third, a drug scan method was utilized to determine the ADME-TOX feasibility of the leads. Last, the molecular dynamics method was applied to determine the chemical reaction repertoire in the function of time. Our designed cyclic-peptide drugs have a much better performance in each of the bioinformatics parameters than other, non-cyclic ones (Tambunan and Parikesit 2013). The penta- and hexa-peptide designs are more preferable than the tripeptide design because the longer the sequences, the more thermodynamically favorable they are. In this regard, this approach should always be considered for developing future drugs against the influenza A virus. The most important aspect of using our molecular simulation tools is the mastery of biochemistry. Without its fundamentals, the data interpretation of the simulation will not be correct or feasible (Adcock and McCammon 2006).

The Future of Influenza A Drug Development: De Novo Molecular Design

It is clear that the lesson learnt from oseltamivir development is that 'rigid' and conventional drug design cannot cope with the challenge of rapid viral mutation. In this regard, novel approaches and methods must keep developing in the scientific community in order to provide the most efficient solution. The problem is the increasing challenges that emerged in this area. The relieving news is that so far there has been no reported person-to-person transmission of the novel influenza A virus (WHO 2017).

More HPAI viruses are emerging, such as H7N9, that the conventional drug cannot cope with (Gao et al. 2013). Thus, to observe the feasibility of a molecular-based drug, gene expression profiling with the assistance of a transcriptomics-based databases such as Ingenuity Pathway Analysis and Connectivity Map would be needed to determine the exact metabolic pathway to inhibit the viral replication (Morrison and Katze 2015). Moreover, the development of system biology that integrates –omics studies for influenza A virus-host interaction is on the way as well (Söderholm et al. 2016). A database that was developed for system biology studies is STRINGdb, which analyzes the protein–protein interaction network in the host cell (Szklarczyk et al. 2017). Figure 5 shows the protein–protein interaction network of neuraminidase proteins in the host cell, where it is clear that it acts synergically with other proteins in the vicinity of the network. Some research has already utilized STRINGdb to observing the protein–protein interaction network in the influenza A virus (Bavagnoli et al. 2011; Ding et al. 2016; Karthick et al. 2013). Thus, there are many potential targets for influenza A drug development besides the conventional ones already discussed.

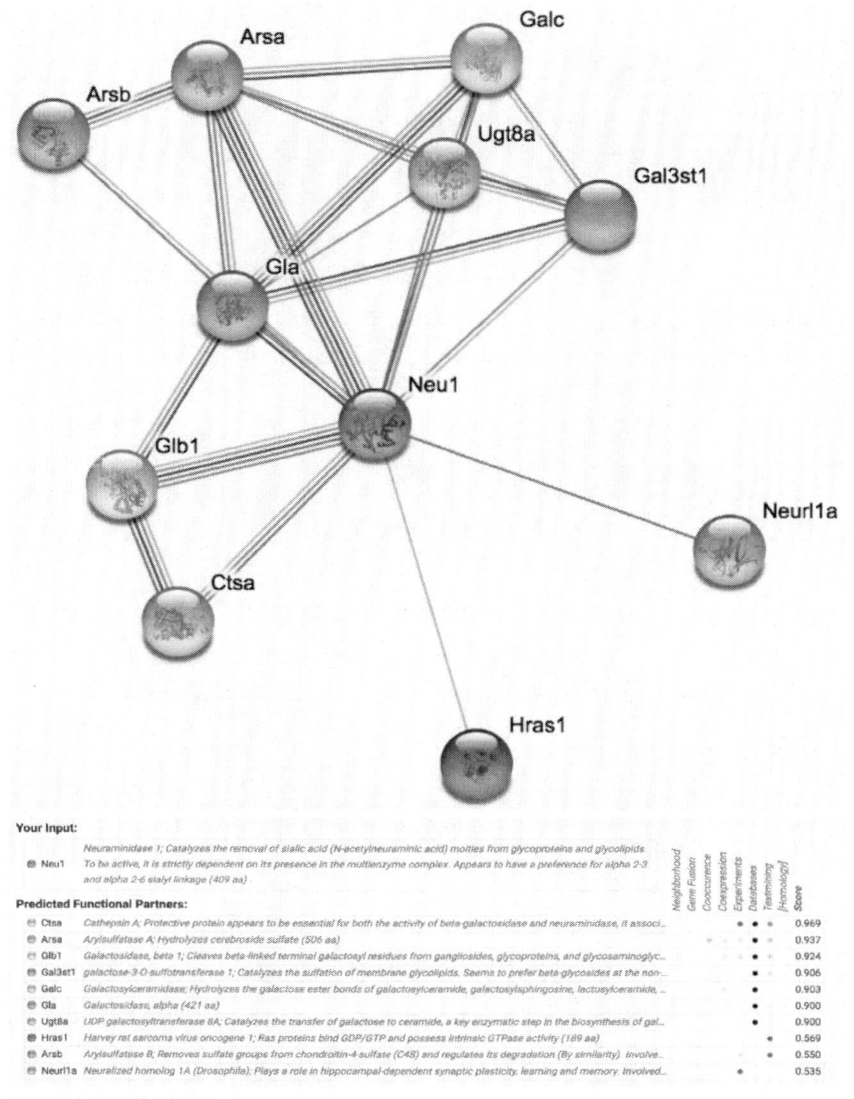

Source: https://string-db.org.

Figure 5. The protein–protein interaction of neuraminidase protein in house mice (Mus musculus) is shown in the upper section, and the legend is in the lower section. It is shown that the neuraminidase enzyme does not act alone in the host cell and works in a synergic manner with other proteins within the network.

Although 'big data'-based -omics studies have made great strides, molecular simulation-based drug development from the structural bioinformatics field should always be considered as well. The study of RNA–protein interactions must be considered, because this part still has not been thoroughly investigated by the scientific community (Lu et al. 2013). The complexity of the interactions should be examined with cutting-edge molecular simulation approaches, which could be strengthened with machine-learning-based methods to devise the clustering of lead compounds in order to find the most potent inhibitors (Li et al. 2016). This approach combines standalone molecular simulation with a 'big data' method to annotate the inhibitors for neuraminidase and other influenza proteins or genes (Zhang et al. 2017).

CONCLUSION

It is concluded that after oseltamivir was excluded from the WHO's 'core drug' list, more viable alternatives that developed using the latest methods in proteomics and transcriptomics studies are giving hope for a remedy to the influenza A infection. On a molecular basis, the development of cyclic-peptide drugs is providing viable alternatives in transcriptomics. While the transcriptomics study provided siRNA therapy as the common ground, in the future, the systematic biological approach that integrates all –omics studies should be devised for giving a more holistic approach to drug design for the influenza A virus. Novel approaches in molecular virology should be intensively developed to protect us against the influenza A virus because nobody knows when the next pandemic events will be. In addition, the possibility of the influenza A virus becoming mutated into a virus that facilitates easy human-to-human transmission cannot be ruled out at all.

ACKNOWLEDGMENTS

The authors would like to thank the Directorate of Research and Community Engagement (DRPM), University of Indonesia, and the

Research and Community Engagements Institute (LPPM), Indonesia International Institute for Life Sciences, for providing heartfelt support for the writing of this manuscript. Thanks also goes to Direktorat Riset dan Pengabdian Masyarakat, Direktorat Jenderal Penguatan Riset dan Pengembangan Kementerian Riset, Teknologi, dan Pendidikan Tinggi Republik Indonesia for providing Hibah Penelitian Berbasis Kompetensi DIKTI/KOPERTIS III 2018 No. 049/KM/PNT/2018. All authors declare that they have no competing interests and that they approved the final version of this manuscript.

REFERENCES

Adcock, Stewart A., and J. Andrew McCammon. 2006. "Molecular Dynamics: Survey of Methods for Simulating the Activity of Proteins." *Chemical Reviews* 106 (5): 1589–1615. https://doi.org/10.1021/cr040426m.

Bavagnoli, Laura, William G. Dundon, Anna Garbelli, Bianca Zecchin, Adelaide Milani, Geetha Parakkal, Fausto Baldanti, et al. 2011. "The PDZ-Ligand and Src-Homology Type 3 Domains of Epidemic Avian Influenza Virus NS1 Protein Modulate Human Src Kinase Activity during Viral Infection." Edited by Andrew Pekosz. *PLoS ONE* 6 (11): e27789. https://doi.org/10.1371/journal.pone.0027789.

Behrens, G., R. Gottschalk, L. Guertler, T.C. Harder, C. Hoffman, B.S. Kamps, S. Korsman, et al. 2006. *Influenza Report*. Edited by Bernd Sebastian Kamps, Christina Hoffman, and Wolfgang Preiser. Sevilla: Flying Publisher. www.influenzareports.com.

Bogdanowich-Knipp, S. J., S. Chakrabarti, T. D. Williams, R. K. Dillman, and T. J. Siahaan. 1999. "Solution Stability of Linear vs. Cyclic RGD Peptides." *The Journal of Peptide Research : Official Journal of the American Peptide Society* 53 (5): 530–41. http://www.ncbi.nlm.nih.gov/pubmed/10424348.

Butcher, Eugene C., Ellen L. Berg, and Eric J. Kunkel. 2004. "Systems Biology in Drug Discovery." *Nature Biotechnology* 22 (10): 1253–59. https://doi.org/10.1038/nbt1017.

Butler, D. 2001. "Genomics. Are You Ready for the Revolution?" *Nature* 409 (6822): 758–60. https://doi.org/10.1038/35057400.

Chandra, Nagasuma, Praveen Anand, and Kalidas Yeturu. 2010. "Structural Bioinformatics: Deriving Biological Insights from Protein Structures." *Interdisciplinary Sciences: Computational Life Sciences* 2 (4): 347–66. https://doi.org/10.1007/s12539-010-0045-6.

Collins, F. S., and V. A. McKusick. 2001. "Implications of the Human Genome Project for Medical Science." *JAMA : The Journal of the American Medical Association* 285 (5): 540–44. http://www.ncbi.nlm.nih.gov/pubmed/11176855.

Davidov, Eugene J., Joanne M. Holland, Edward W. Marple, and Stephen Naylor. 2003. "Advancing Drug Discovery through Systems Biology." *Drug Discovery Today* 8 (4): 175–83. https://doi.org/10.1016/S1359-6446(03)02600-X.

Davis, L. E. 2014. "Influenza Virus." In *Encyclopedia of the Neurological Sciences*, 36:695–97. Karger Publishers. https://doi.org/10.1016/B978-0-12-385157-4.00381-X.

Day, Joshua A., and Seth M. Cohen. 2013. "Investigating the Selectivity of Metalloenzyme Inhibitors." *Journal of Medicinal Chemistry* 56 (20): 7997–8007. https://doi.org/10.1021/jm401053m.

Desany, Brian, and Zemin Zhang. 2004. "Bioinformatics and Cancer Target Discovery." *Drug Discovery Today*. https://doi.org/10.1016/S1359-6446(04)03224-6.

Ding, Xiaoman, Jiahai Lu, Ruoxi Yu, Xin Wang, Ting Wang, Fangyuan Dong, Bo Peng, et al. 2016. "Preliminary Proteomic Analysis of A549 Cells Infected with Avian Influenza Virus H7N9 and Influenza a Virus H1N1." Edited by Suryaprakash Sambhara. *PLoS ONE* 11 (5): e0156017. https://doi.org/10.1371/journal.pone.0156017.

Dong, Zhicheng, and Yan Chen. 2013. "Transcriptomics: Advances and Approaches." *Science China. Life Sciences* 56 (10): 960–67. https://doi.org/10.1007/s11427-013-4557-2.

Drugbank. 2018. "Oseltamivir Entry in Drugbank." *Drugbank.* 2018. https://www.drugbank.ca/drugs/DB00198.

Dykxhoorn, Derek M., and Judy Lieberman. 2005. "The Silent Revolution: RNA Interference as Basic Biology, Research Tool, and Therapeutic." *Annual Review of Medicine* 56 (1): 401–23. https://doi.org/10.1146/annurev.med.56.082103.104606.

Edwards, Richard J., Niamh Moran, Marc Devocelle, Aoife Kiernan, Gerardene Meade, William Signac, Martina Foy, et al. 2007. "Bioinformatic Discovery of Novel Bioactive Peptides." *Nature Chemical Biology* 3 (2): 108–12. https://doi.org/10.1038/nchembio854.

Ferraris, Olivier, and Bruno Lina. 2008. "Mutations of Neuraminidase Implicated in Neuraminidase Inhibitors Resistance." *Journal of Clinical Virology : The Official Publication of the Pan American Society for Clinical Virology* 41 (1): 13–19. https://doi.org/10.1016/j.jcv.2007.10.020.

Gao, Rongbao, Bin Cao, Yunwen Hu, Zijian Feng, Dayan Wang, Wanfu Hu, Jian Chen, et al. 2013. "Human Infection with a Novel Avian-Origin Influenza A (H7N9) Virus." *The New England Journal of Medicine* 368 (20): 1888–97. https://doi.org/10.1056/NEJMoa1304459.

Gomase, V. 2009. *Transcriptomics: Expression Pattern Analysis.* Lightning Source Incorporated. http://books.google.co.id/books?id=1yR2QgAACAAJ.

Grienke, Ulrike, Michaela Schmidtke, Susanne von Grafenstein, Johannes Kirchmair, Klaus R. Liedl, and Judith M. Rollinger. 2012. "Influenza Neuraminidase: A Druggable Target for Natural Products." *Natural Product Reports* 29 (1): 11–36. https://doi.org/10.1039/c1np00053e.

Gunsteren, Wilfred F. van, Dirk Bakowies, Riccardo Baron, Indira Chandrasekhar, Markus Christen, Xavier Daura, Peter Gee, et al. 2006. "Biomolecular Modeling: Goals, Problems, Perspectives." *Angewandte Chemie (International Ed. in English)* 45 (25): 4064–92. https://doi.org/10.1002/anie.200502655.

Hama, Rokuro. 2015. "Clinical Microbiology : Open Access The Mechanisms of Adverse Reactions to Oseltamivir : Part II. Delayed

Type Reactions." *Clinical Microbiology: Open Access* 4 (6): 1–10. https://doi.org/10.4172/2327-5073.1000224.

Hannon, G. J. 2002. "RNA Interference." *Nature* 418 (6894): 244–51. https://doi.org/10.1038/418244a [doi]\r418244a [pii].

Hiasa, Miki, Yumiko Isoda, Yasushi Kishimoto, Kenta Saitoh, Yasuaki Kimura, Motomu Kanai, Masakatsu Shibasaki, Dai Hatakeyama, Yutaka Kirino, and Takashi Kuzuhara. 2013. "Inhibition of MAO-A and Stimulation of Behavioural Activities in Mice by the Inactive Prodrug Form of the Anti-Influenza Agent Oseltamivir." *British Journal of Pharmacology*. https://doi.org/10.1111/bph.12102.

Hsu, Peng-Hao, Din-Chi Chiu, Kuan-Lin Wu, Pei-Shan Lee, Jia-Tsrong Jan, Yih-Shyun E. Cheng, Keng-Chang Tsai, Ting-Jen Cheng, and Jim-Min Fang. 2018. "Acylguanidine Derivatives of Zanamivir and Oseltamivir: Potential Orally Available Prodrugs against Influenza Viruses." *European Journal of Medicinal Chemistry* 154 (June): 314–23. https://doi.org/10.1016/j.ejmech.2018.05.030.

Huang, Hung-Jin Jin, Hsin Wei Yu, Chien Yu Calvin Yu Chian Chien-Yu Calvin Yu-Chian Chian Chien Yu Chen, Chih-Ho Ho Hsu, Hsin-Yi Yi Chen, Kuei-Jen Jen Lee, Fuu-Jen Jen Tsai, and Chien Yu Calvin Yu Chian Chien-Yu Calvin Yu-Chian Chian Chien Yu Chen. 2010. "Current Developments of Computer-Aided Drug Design." *Journal of the Taiwan Institute of Chemical Engineers* 41 (6): 623–35. https://doi.org/10.1016/j.jtice.2010.03.017.

Iskandarsyah, Bimo A. Tejo, Usman S. F. Tambunan, Gennady Verkhivker, and Teruna J. Siahaan. 2008. "Structural Modifications of ICAM-1 Cyclic Peptides to Improve the Activity to Inhibit Heterotypic Adhesion of T Cells." *Chemical Biology & Drug Design* 72 (1): 27–33. https://doi.org/10.1111/j.1747-0285.2008.00676.x.

Jónsdóttir, Svava Osk, Flemming Steen Jørgensen, and Søren Brunak. 2005. "Prediction Methods and Databases within Chemoinformatics: Emphasis on Drugs and Drug Candidates." *Bioinformatics (Oxford, England)* 21 (10): 2145–60. https://doi.org/10.1093/bioinformatics/bti314.

Kandun, I. Nyoman, Hariadi Wibisono, Endang R. Sedyaningsih, Yusharmen, Widarso Hadisoedarsuno, Wilfried Purba, Hari Santoso, et al. 2006. “Three Indonesian Clusters of H5N1 Virus Infection in 2005.” *New England Journal of Medicine* 355 (21): 2186–94. https://doi.org/10.1056/NEJMoa060930.

Karthick, V., K. Ramanathan, V. Shanthi, and R. Rajasekaran. 2013. “Identification of Potential Inhibitors of H5N1 Influenza A Virus Neuraminidase by Ligand-Based Virtual Screening Approach.” *Cell Biochemistry and Biophysics* 66 (3): 657–69. https://doi.org/10.1007/s12013-012-9510-7.

Kemenkes-RI. 2013. “Profil Kesehatan Indonesia.” *Jakarta.* http://www.depkes.go.id/resources/download/pusdatin/profil-kesehatan-indonesia/profil-kesehatan-indonesia-2013.pdf.

Kmietowicz, Zosia. 2017. “WHO Downgrades Oseltamivir on Drugs List after Reviewing Evidence.” *BMJ (Clinical Research Ed.)* 357 (June): j2841. https://doi.org/10.1136/bmj.j2841.

Kocik, Janusz, Marcin Kołodziej, Justyna Joniec, Magdalena Kwiatek, and Michał Bartoszcze. 2014. “Antiviral Activity of Novel Oseltamivir Derivatives against Some Influenza Virus Strains.” *Acta Biochimica Polonica* 61 (3): 509–13. http://www.ncbi.nlm.nih.gov/pubmed/25210935.

Kumari, Indu, Padmani Sandhu, Mushtaq Ahmed, and Yusuf Akhter. 2017. “Molecular Dynamics Simulations, Challenges and Opportunities: A Biologist’s Prospective.” *Current Protein & Peptide Science* 18 (11): 1163–79. https://doi.org/10.2174/1389203718666170622074741.

Li, Rui, Yeng-Tseng Wang, and Cheng-Lung Chen. 2013. “Computational Modeling Study on Metabolism Mechanism of Oseltamivir.” *Journal of Theoretical and Computational Chemistry* 12 (05): 1350037. https://doi.org/10.1142/S0219633613500375.

Li, Yang, Yue Kong, Mengdi Zhang, Aixia Yan, and Zhenming Liu. 2016. “Using Support Vector Machine (SVM) for Classification of Selectivity of H1N1 Neuraminidase Inhibitors.” *Molecular Informatics* 35 (3–4): 116–24. https://doi.org/10.1002/minf.201500107.

Lindemann, Lothar, Helmut Jacobsen, Diana Schuhbauer, Frederic Knoflach, Silvia Gatti, Joseph G. Wettstein, Hansruedi Loetscher, et al. 2010. "In Vitro Pharmacological Selectivity Profile of Oseltamivir Prodrug (Tamiflu®) and Active Metabolite." *European Journal of Pharmacology*. https://doi.org/10.1016/j.ejphar.2009.11.020.

Lu, Qiongshi, Sijin Ren, Ming Lu, Yong Zhang, Dahai Zhu, Xuegong Zhang, and Tingting Li. 2013. "Computational Prediction of Associations between Long Non-Coding RNAs and Proteins." *BMC Genomics*. https://doi.org/10.1186/1471-2164-14-651.

Luzhkov, Victor B., Barbara Selisko, Anneli Nordqvist, Frédéric Peyrane, Etienne Decroly, Karine Alvarez, Anders Karlen, Bruno Canard, and Johan Qvist. 2007. "Virtual Screening and Bioassay Study of Novel Inhibitors for Dengue Virus MRNA Cap (Nucleoside-2'O)-Methyltransferase." *Bioorganic & Medicinal Chemistry* 15 (24): 7795–7802. https://doi.org/10.1016/j.bmc.2007.08.049.

MacKerell, A. D., D. Bashford, R. L. Dunbrack, J. D. Evanseck, M. J. Field, S. Fischer, J. Gao, et al. 1998. "All-Atom Empirical Potential for Molecular Modeling and Dynamics Studies of Proteins." *The Journal of Physical Chemistry B* 102 (18): 3586–3616. https://doi.org/10.1021/jp973084f.

Marz, Manja, Niko Beerenwinkel, Christian Drosten, Markus Fricke, Dmitrij Frishman, Ivo L. Hofacker, Dieter Hoffmann, et al. 2014. "Challenges in RNA Virus Bioinformatics." *Bioinformatics* 30 (13): 1793–99. https://doi.org/10.1093/bioinformatics/btu105.

Morrison, Juliet, and Michael G. Katze. 2015. "Gene Expression Signatures as a Therapeutic Target for Severe H7N9 Influenza – What Do We Know so Far?" *Expert Opinion on Therapeutic Targets* 19 (4): 447–50. https://doi.org/10.1517/14728222.2015.1006198.

Muraki, Katsuhiko, Noriyuki Hatano, Hiroka Suzuki, Yukiko Muraki, Yui Iwajima, Yasuhiro Maeda, and Hideki Ono. 2015. "Oseltamivir Blocks Human Neuronal Nicotinic Acetylcholine Receptor-Mediated Currents." *Basic and Clinical Pharmacology and Toxicology*. https://doi.org/10.1111/bcpt.12290.

Nakamura, Haruki. 2007. "Proteomics Research at the Institute for Protein Research, Osaka University." *Asia-Pacific Biotech News* 11 (15): 1040–44. https://doi.org/10.1142/S0219030307001103.

Naulet, Guillaume, Antoine Robert, Pierre Dechambenoit, Harald Bock, and Fabien Durola. 2018. "Monoprotection of Arylene-Diacetic Acids Allowing the Build-Up of Longer Aromatic Ribbons by Successive Perkin Condensations." *European Journal of Organic Chemistry* 2018 (5): 619–26. https://doi.org/10.1002/ejoc.201701499.

On, Ngoc H., Paul Kiptoo, Teruna J. Siahaan, and Donald W. Miller. 2014. "Modulation of Blood-Brain Barrier Permeability in Mice Using Synthetic E-Cadherin Peptide." *Molecular Pharmaceutics*, February. https://doi.org/10.1021/mp400624v.

Parikesit, A. A., B. Ardiansah, D. M. Handayani, U. S. F. Tambunan, and D. Kerami. 2016. "Virtual Screening of Indonesian Flavonoid as Neuraminidase Inhibitor of Influenza a Subtype H5N1." *IOP Conference Series: Materials Science and Engineering* 107 (1): 012053. https://doi.org/10.1088/1757-899X/107/1/012053.

Parikesit, Arli Aditya. 2009. "The Role of Bioinformatics as Auxilliary Tools for Molecular Biology." In *Proceeding of World-Wide Indonesian Student Association Scientific Writing Olympic*, 23–29. Paris: Indonesian Student Association of France. http://www.vixra.org/abs/1308.0115.

Parikesit, Arli Aditya. 2018. "The Construction of Two and Three Dimensional Molecular Models for the MiR-31 and Its Silencer as the Triple Negative Breast Cancer Biomarkers." *OnLine Journal of Biological Sciences*, October. https://doi.org/10.3844/OFSP.12140.

Parikesit, Arli Aditya, Harry Noviardi, Djati Kerami, and Usman Sumo Friend Tambunan. 2014. "The Complexity of Molecular Interactions and Bindings between Cyclic Peptide and Inhibit Polymerase A and B1 (PAC-PB1N) H1N1." In *9th Joint Conference on Chemistry*, 1:382–85. Semarang: UNNES Press. https://doi.org/10.13140/RG.2.1.1439.6969.

Parikesit, Arli Aditya, Didik Huswo Utomo, and Nihayatul Karimah. 2018. "Determination of Secondary and Tertiary Structures of Cervical Cancer LncRNA Diagnostic and SiRNA Therapeutic Biomarkers." *Indonesian*

Journal of Biotechnology 23 (1): 1. https://doi.org/10.22146/ijbiotech.28508.

Pickett, Brett E., Eva L. Sadat, Yun Zhang, Jyothi M Noronha, R. Burke Squires, Victoria Hunt, Mengya Liu, et al. 2012. "ViPR: An Open Bioinformatics Database and Analysis Resource for Virology Research." *Nucleic Acids Research* 40 (Database issue): D593-8. https://doi.org/10.1093/nar/gkr859.

Pubchem. 2018. "Oseltamivir Entry in PUBCHEM." *Pubchem.* 2018. https://pubchem.ncbi.nlm.nih.gov/compound/oseltamivir#section=Top.

Russell, Rupert J., Lesley F. Haire, David J. Stevens, Patrick J. Collins, Yi Pu Lin, G. Michael Blackburn, Alan J. Hay, Steven J. Gamblin, and John J. Skehel. 2006. "The Structure of H5N1 Avian Influenza Neuraminidase Suggests New Opportunities for Drug Design." *Nature* 443 (7107): 45–49. https://doi.org/10.1038/nature05114.

Sætrom, Pål, and Ola Snøve. 2004. "A Comparison of SiRNA Efficacy Predictors." *Biochemical and Biophysical Research Communications* 321 (1): 247–53. https://doi.org/10.1016/J.BBRC.2004.06.116.

Sattar, Rabia, Syed Abid Ali, Mustafa Kamal, Aftab Ahmed Khan, and Atiya Abbasi. 2004. "Molecular Mechanism of Enzyme Inhibition: Prediction of the Three-Dimensional Structure of the Dimeric Trypsin Inhibitor from Leucaena Leucocephala by Homology Modelling." *Biochemical and Biophysical Research Communications* 314 (3): 755–65. https://doi.org/10.1016/j.bbrc.2003.12.177.

Shaikh, S. A., T. Jain, G. Sandhu, N. Latha, and B. Jayaram. 2007. "From Drug Target to Leads--Sketching a Physicochemical Pathway for Lead Molecule Design in Silico." *Current Pharmaceutical Design* 13: 3454–70. https://doi.org/10.2174/138161207782794220.

Shaji, Jessy, and V. Patole. 2008. "Protein and Peptide Drug Delivery: Oral Approaches." *Indian Journal of Pharmaceutical Sciences* 70 (3): 269–77. https://doi.org/10.4103/0250-474X.42967.

Shi, Deshi, J. Yang, Dongfang Yang, Edward L. LeCluyse, Chris Black, Li You, Fatemeh Akhlaghi, and Bingfang Yan. 2006. "Anti-Influenza Prodrug Oseltamivir Is Activated by Carboxylesterase Human Carboxylesterase 1, and the Activation Is Inhibited by Antiplatelet

Agent Clopidogrel." *J Pharmacol Exp Ther* 319 (3): 1477–84. https://doi.org/10.1124/jpet.106.111807.

Shtyrya, Y. A., L. V. Mochalova, and N. V. Bovin. 2009. "Influenza Virus Neuraminidase: Structure and Function." *Acta Naturae* 1 (2): 26–32. http://www.ncbi.nlm.nih.gov/pubmed/22649600.

Sinaga, Ernawati, Seetharama D. S. Jois, Mike Avery, Irwan T. Makagiansar, Usman S. F. Tambunan, Kenneth L. Audus, and Teruna J. Siahaan. 2002. "Increasing Paracellular Porosity by E-Cadherin Peptides: Discovery of Bulge and Groove Regions in the EC1-Domain of E-Cadherin." *Pharmaceutical Research* 19 (8): 1170–79. http://www.ncbi.nlm.nih.gov/pubmed/12240943.

Söderholm, Sandra, Yu Fu, Lana Gaelings, Sergey Belanov, Laxman Yetukuri, Mikhail Berlinkov, Anton V. Cheltsov, et al. 2016. "Multi-Omics Studies towards Novel Modulators of Influenza a Virus-Host Interaction." *Viruses*. Multidisciplinary Digital Publishing Institute. https://doi.org/10.3390/v8100269.

Soltero, Rick. 2005. "Oral Protein and Peptide Drug Delivery." In *Drug Delivery: Principles and Application*, edited by Binghe Wang, Teruna J. Siahaan, and Richard Soltero, 189–200. John Wiley & Sons, Inc.

Spessard, G.O. 1998. "ACD Labs/LogP DB 3.5 and ChemSketch 3.5." *Journal of Chemical Information and Modeling* 38 (6): 1250–53. https://doi.org/10.1021/ci980264t.

Sugrue, Richard J., Boon Huan Tan, Dawn S. Y. Yeo, and Richard Sutejo. 2008. "Antiviral Drugs for the Control of Pandemic Influenza Virus." *Annals of the Academy of Medicine, Singapore* 37 (6): 518–24. http://www.ncbi.nlm.nih.gov/pubmed/18618065.

Syahdi, Rezi Riadhi, Abdul Mun'im, Heru Suhartanto, and Arry Yanuar. 2012. "Virtual Screening of Indonesian Herbal Database as HIV-1 Reverse Transcriptase Inhibitor." *Bioinformation* 8 (24): 1206–10. https://doi.org/10.6026/97320630081206.

Szklarczyk, Damian, John H. Morris, Helen Cook, Michael Kuhn, Stefan Wyder, Milan Simonovic, Alberto Santos, et al. 2017. "The STRING Database in 2017: Quality-Controlled Protein–Protein Association

Networks, Made Broadly Accessible." *Nucleic Acids Research* 45 (D1): D362–68. https://doi.org/10.1093/nar/gkw937.

Takano, Ryo, Maki Kiso, Manabu Igarashi, Quynh Mai Le, Masakazu Sekijima, Kimihito Ito, Ayato Takada, and Yoshihiro Kawaoka. 2013. "Molecular Mechanisms Underlying Oseltamivir Resistance Mediated by an I117V Substitution in the Neuraminidase of Subtype H5N1 Avian Influenza a Viruses." *Journal of Infectious Diseases* 207 (1): 89–97. https://doi.org/10.1093/infdis/jis633.

Tambunan, Usman Sumo Friend., Noval Amri, and Arli Aditya Parikesit. 2012. "In Silico Design of Cyclic Peptides as Influenza Virus, a Subtype H1N1 Neuraminidase Inhibitor." *African Journal of Biotechnology* 11 (52): 11474–91. https://doi.org/10.5897/AJB11.4094.

Tambunan, Usman Sumo Friend, Mochammad Arfin Fardiansyah Nasution, Arli Aditya Parikesit, Harry Noviardi, and Djati Kerami. 2016. "Designing of Disulfide Cyclic Peptide for Inhibiting Polymerase A and B1 (PAC-PB1N) in H1N1 Virus Using Molecular Simulation Approach." *OnLine Journal of Biological Sciences* 16 (3): 122–29. https://doi.org/10.3844/ojbsci.2016.122.129.

Tambunan, Usman Sumo Friend, and Arli Aditya Parikesit. 2010. "Cracking the Genetic Code of Human Virus by Using Open Source Bioinformatics Tools." *Malaysian Journal of Fundamental and Applied Sciences* 6 (1): 42–50. https://doi.org/10.11113/mjfas.v6n1.175.

Tambunan, Usman Sumo Friend, and Arli Aditya Parikesit. 2013. "In Silico Approach towards H5N1 Virus Protein and Transcriptomics-Based Medication." *African Journal of Biotechnology* 12 (21): 3110–14. https://doi.org/10.5897/AJBX12.013.

Tambunan, Usman Sumo Friend, Arli Aditya Parikesit, Yonaniko Dephinto, and Feimmy Ruth Pratiwi Sipahutar. 2014. "Computational Design of Drug Candidates for Influenza A Virus Subtype H1N1 by Inhibiting the Viral Neuraminidase-1 Enzyme." *Acta Pharmaceutica* 64 (2): 157–72. https://doi.org/10.2478/acph-2014-0015.

Tambunan, Usman Sumo Friend, Arli Aditya Parikesit, Yossy Carolina Unadi, and Djati Kerami. 2015. "Designing Cyclopentapeptide Inhibitor of Neuraminidase H5N1 Virus Through Molecular and Pharmacology

Simulations." *Tsinghua Science and Technology* 20 (5): 431–40. https://doi.org/10.1109/TST.2015.7297742.

Taubenberger, Jeffery K, and David M Morens. 2006. "1918 Influenza: The Mother of All Pandemics." *Emerging Infectious Diseases* 12 (1): 15–22. https://doi.org/10.3201/eid1201.050979.

Uhlén, Mathias, Linn Fagerberg, Björn M Hallström, Cecilia Lindskog, Per Oksvold, Adil Mardinoglu, Åsa Sivertsson, et al. 2015. "Proteomics. Tissue-Based Map of the Human Proteome." *Science (New York, N.Y.)*. https://doi.org/10.1126/science.1260419.

Ungchusak, Kumnuan, Prasert Auewarakul, Scott F. Dowell, Rungrueng Kitphati, Wattana Auwanit, Pilaipan Puthavathana, Mongkol Uiprasertkul, et al. 2005. "Probable Person-to-Person Transmission of Avian Influenza A (H5N1)." *New England Journal of Medicine* 352 (4): 333–40. https://doi.org/10.1056/NEJMoa044021.

Vavricka, Christopher J., Qing Li, Yan Wu, Jianxun Qi, Mingyang Wang, Yue Liu, Feng Gao, et al. 2011. "Structural and Functional Analysis of Laninamivir and Its Octanoate Prodrug Reveals Group Specific Mechanisms for Influenza NA Inhibition." *PLoS Pathogens* 7 (10): e1002249. https://doi.org/10.1371/journal.ppat.1002249.

Welch, Lonnie, Fran Lewitter, Russell Schwartz, Cath Brooksbank, Predrag Radivojac, Bruno Gaeta, and Maria Victoria Schneider. 2014. "Bioinformatics Curriculum Guidelines: Toward a Definition of Core Competencies." *PLoS Computational Biology* 10 (3): e1003496. https://doi.org/10.1371/journal.pcbi.1003496.

Wheeler, David L., Tanya Barrett, Dennis A. Benson, Stephen H. Bryant, Kathi Canese, Vyacheslav Chetvernin, Deanna M. Church, et al. 2007. "Database Resources of the National Center for Biotechnology Information." *Nucleic Acids Research* 35 (Database issue): D5-12. https://doi.org/10.1093/nar/gkl1031.

Whittaker, Paul A. 2003. "What Is the Relevance of Bioinformatics to Pharmacology?" *Trends in Pharmacological Sciences*. https://doi.org/10.1016/S0165-6147(03)00197-4.

WHO. 2010. "WHO | Oseltamivir (Inclusion)." *WHO*. https://www.who.int/selection_medicines/committees/expert/17/application/oseltamivir/en/.

WHO. 2012. "Situation Updates - Avian Influenza," 6. http://www.who.int/influenza/human_animal_interface/en/.

WHO. 2014. "Cumulative Number of Confirmed Human Cases for Avian Influenza A (H5N1) Reported to WHO."

WHO. 2015. "WHO | What Is a Pandemic?" *WHO*. https://www.who.int/csr/disease/swineflu/frequently_asked_questions/pandemic/en/.

WHO. 2017. "WHO | Avian Influenza A(H7N9) Virus." *WHO*. https://www.who.int/influenza/human_animal_interface/influenza_h7n9/en/.

Winter, Greg, and Stan Fields. 1982. "Nucteotide Sequence of Human Influenza A/PR/8/34 Segment 2." *Nucleic Acids Research* 10 (6): 2135–43. https://doi.org/10.1093/nar/10.6.2135.

Wishart, David S., Craig Knox, An Chi Guo, Dean Cheng, Savita Shrivastava, Dan Tzur, Bijaya Gautam, and Murtaza Hassanali. 2008. "DrugBank: A Knowledgebase for Drugs, Drug Actions and Drug Targets." *Nucleic Acids Research* 36 (Database issue): D901–6.

Wu, Xiaogang, Mohammad Al Hasan, and Jake Yue Chen. 2014. "Pathway and Network Analysis in Proteomics." *Journal of Theoretical Biology* 362: 44–52. https://doi.org/10.1016/j.jtbi.2014.05.031.

Wu, Yan, Yuhai Bi, Christopher J Vavricka, Xiaoman Sun, Yanfang Zhang, Feng Gao, Min Zhao, et al. 2013. "Characterization of Two Distinct Neuraminidases from Avian-Origin Human-Infecting H7N9 Influenza Viruses." *Cell Research* 23 (12): 1347–55. https://doi.org/10.1038/cr.2013.144.

Yagi, Yukiko, Kotaro Terada, Takahisa Noma, Kazunori Ikebukuro, and Koji Sode. 2007. "In Silico Panning for a Non-Competitive Peptide Inhibitor." *BMC Bioinformatics* 8 (1): 11. https://doi.org/10.1186/1471-2105-8-11.

Yanuar, Arry, Abdul Mun'im, Akma Bertha Aprima Lagho, Rezi Riadhi Syahdi, Marjuqi Rahmat, and Heru Suhartanto. 2011. "Medicinal Plants Database and Three Dimensional Structure of the Chemical Compounds from Medicinal Plants in Indonesia." Biomolecules. *International Journal of Computer Science* 8 (5): 180–83. http://arxiv.org/abs/1111.7183.

Yanuar, Arry, Heru Suhartanto, Abdul Mun'im, Bram Hik Anugraha, and Rezi Riadhi Syahdi. 2014. "Virtual Screening of Indonesian Herbal Database as HIV-1 Protease Inhibitor." *Bioinformation* 10 (2): 52–55. http://www.bioinformation.net/010/97320630010052.pdf.

Zhang, Li, Haixin Ai, Qi Zhao, Junfeng Zhu, Wen Chen, Xuewei Wu, Liangchao Huang, Zimo Yin, Jian Zhao, and Hongsheng Liu. 2017. *Computational Prediction of Influenza Neuraminidase Inhibitors Using Machine Learning Algorithms and Recursive Feature Elimination Method.* In, 344–49. Springer, Cham. https://doi.org/10.1007/978-3-319-59575-7_32.

In: Proteomics
Editor: Ricardo Parker

ISBN: 978-1-53616-440-4

Chapter 3

MASS SPECTROMETRY-BASED PROTEOMICS FOR PRETERM LABOR

***Danai Mavreli*[1,2], *Nikolas Papantoniou*[1], *Dimitrios Gourgiotis*[3] and *Aggeliki Kolialexi*[1,2,*]**

[1]3rd Department of Obstetrics and Gynecology, National and Kapodistrian University of Athens Medical School, Athens, Greece

[2]Department of Medical Genetics, National and Kapodistrian University of Athens Medical School, Athens, Greece

[3]Biochemistry and Molecular Biology Unit, 2nd Department of Pediatrics, National and Kapodistrian University of Athens Medical School, Athens, Greece

ABSTRACT

Preterm labor (PTL), defined as delivery <37 weeks of gestation, is a major health issue for modern obstetrics. Approximately 15 million infants

[*] Corresponding Author's E-mail: akolial@med.uoa.gr.

are born prematurely worldwide every year and out of those more than 1 million do not reach the second year of life. Preterm infants are also at high risk of long-term adverse effects.

The etiology remains unclear but it is likely multifactorial, linked to impaired placental function. Recent studies focusing on placental and maternal peripheral blood protein profiling using mass spectrometry-based proteomic technology show differential expression between normal and complicated pregnancies, providing valuable information about the pathophysiological role of proteins and identifying potential biomarkers for monitoring pregnancy complications.

Herein we review the most recent research studies aiming to identify biomarkers for the early prediction of sPTL using mass spectrometry-based proteomics.

INTRODUCTION

Preterm labor (PTL), defined as delivery before 37 weeks of gestation, is one of the most critical issues modern obstetrics is facing with a prevalence of 11% worldwide [1]. Clinically, PTL is described as either spontaneous PTL (sPTL) accounting for approximately 75% of preterm deliveries or iatrogenic, mainly attributed to preeclampsia or fetal growth restriction (IUGR) [2].

SPTL is the leading cause of neonatal mortality and morbidity accounting for 75% of all perinatal deaths and the second most frequent cause of death in children aged <5 years old [3]. Premature infants who do survive, require prolonged hospitalization and, compared to those born at term, are at high risk for respiratory distress, apnea, hypoglycemia, necrotizing enterocolitis and seizures [4]. Born at pre-term, many infants face life-long disabilities including chronic lung disease, hearing impairment and visual neurodevelopmental deficits [5]. The 2015 Global Burden of Disease study ranked PTL as the fifth leading cause of disease burden over time [6]. PTL adds an incredible emotional strain to the families but also presents a significant challenge to society.

While improvements in neonatal care have led to higher survival rates of premature infants, the number of sPTL has annually increased by around 2% over the last 20 years, mainly because of failure to identify pregnancies at high risk for this complication [7]. Accurate prediction of sPTL risk, before the clinical event, may, through effective treatments, either prevent premature delivery or prolong gestation enough to optimize the outcome for the fetus. It is noteworthy that effective preventive treatments, such as cervical cerclage and prophylactic progesterone administration, are available and their effectiveness is currently being tested in various randomized trials [8-12].

SPTL is a heterogeneous condition, with multiple underlying etiologies. Associated risk factors include a family history of PTL, maternal age, African-American race and smoking [13]. These risk factors, however, do not define the cause of sPTL but merely indicate women who are at increased risk for developing this complication. The strongest predictor is a history of sPTL, which is associated with a 32% chance of recurrent PTL [14]. The majority of sPTL cases, however, occur in women without a prior history, particularly in women in their first pregnancy, highlighting the urgent need for greater international attention on developing improved methods for early prediction of premature births [15]. Despite intense research, up until now, no single biomarker or a combination of biomarkers exists, showing the appropriate specificity and sensitivity to be incorporated into clinical practice [16]. Thus, screening for the early prediction of sPTL remains challenging. Presently, at a 10% false positive rate (FPR), an algorithm combining maternal characteristics and obstetric history, including maternal age, maternal Body Mass Index (BMI), racial origin, smoking status, spontaneous or assisted conception and prior history of sPTL, has been shown to predict at 11-13 weeks of gestation only 18% of nulliparous and 38% of parous women delivering prematurely [17]. The performance of early screening for sPTL has not improved by the addition of a series of biochemical markers, such as maternal serum concentration of pregnancy-associated plasma protein-A (PAPP-A), free β-human chorionic gonadotrophin (β-hCG), placental growth factor (PlGF), placental protein 13 (PP13), A disintegrin and metalloprotease 12 (ADAM12), alpha fetoprotein (aFP), inhibin- A or activin-A [4, 18]. Biophysical markers have also been tested for sPTL screening. Measurement of endocervical length (CL), using transvaginal ultrasound, is one the best predictors during the 2nd trimester of pregnancy [19]. The predictive power, however, during the 1st trimester, when combined with maternal characteristics and obstetric history, was reported to be 54.8%, far below the desired rate for prenatal screening [20-22]. Given the substantial personal, economic and health impact of premature delivery, more accurate predictive tests are needed to highlight those at higher risk for premature delivery who may benefit from preventive interventions.

In the post-genomics era, mass spectrometry-based proteomics has been recognized as the most promising platform for disease-specific biomarker detection in a high-throughput fashion providing insights into the disease-associated pathways.

Herein, we review the literature published on research studies aiming to identify biomarkers for the early prediction of sPTL using mass spectrometry-based proteomics.

PROTEOMIC TECHNOLOGIES

The progress in proteomic technologies in the past decade is mainly attributed to the emergence of Mass Spectrometry-based technology (MS) that allows for the simultaneous identification and quantitation of thousands of proteins in complex samples thus providing improved and novel strategies for the global understanding of cellular functions, for organ biology comprehension and systems biology studies [23]. The advantages of MS-based techniques include sensitivity, resolution, speed and high throughput. They are however time-consuming since data obtained need further analysis using sophisticated bioinformatics tools.

The choice of the MS-based strategy is strongly dependent on the biological question to be addressed and also on high-throughput facilities available to each research group. MS based proteomic strategies are classified as top-down and bottom-up approaches:

Top-down proteomic approaches, the most widely accepted, rely upon a coordinated use of 2-DE analysis coupled with MS. Proteins are first separated using a combination of isoelectric focusing and sodium dodecyl sulfate (SDS) electrophoresis and next visualized either with staining for the visible range (colloidal Coomassie blue) or fluorescence staining. 2D gel electrophoresis has limited sensitivity and is ineffective when it comes to the analysis of low abundance or high molecular weight proteins as well as hydrophobic ones requiring many replicate experiments and background normalization to substantiate in-experiment reproducibility [24-26].

In contrast to gel-based proteomic approaches, *bottom-up proteomics also known as shotgun proteomics* are based on high-resolution liquid chromatography. At the beginning of the workflow, peptides are generated by enzyme digestion and then subjected to high-performance liquid chromatography (HPLC) hyphenated to a tandem mass spectrometer. Shotgun methods may also use label-based techniques for differential tagging of proteins including stable isotope labeling by amino acids in cell culture (SILAC), isotope-coded affinity tag (ICAT), isobaric tags for relative and absolute quantification (iTRAQ) and tandem mass targets (TMT) [27, 28]. Shotgun proteomics provides a two-order of magnitude greater amount of results as compared to gel-based proteomics. They display however low quantification accuracy suggesting that results need further verification by targeted proteomics and/or biochemical assays.

For targeted proteomics, a selected reaction monitoring (SRM) system, measuring the transition of an ion before and after fragmentation, is applied [23]. This approach is more sensitive and specific than shotgun proteomics methods, and thus has the characteristics required for the verification of the data acquired by shotgun proteomics [29].

PROTEOMICS FOR THE IDENTIFICATION OF BIOMARKERS FOR SPTL

Proteomic methodologies have been heavily used for the analysis of biological fluids and tissues related to human reproduction including placental membranes, amniotic fluid (AF), cervical vaginal fluid (CVF) and follicular fluid [30-32]. Several studies have been performed to characterize the human AF proteome but also to study the aberrant expression of specific proteins and identify potential biomarkers in pathological conditions including pregnancy-related complications such as sPTL [25, 33-36]. Use of SELDI-TOF analysis coupled with LC-MS/MS for AF profiling revealed differences in the peak intensity in AF samples obtained from women with PTL and sub-clinical intra-amniotic infection (AIA) [33]. The results indicated calgranulin B and proteolytic fragment IGFBP-1 as potential biomarkers. In another study, Pereira *et al.,* compared the CVF proteomic profile from women who experienced sPTL to that of uncomplicated term pregnancies using DIGE and spectral counting for protein quantitation and found several proteins showing altered expression between the two groups [37]. Butt et al. analyzed placental tissues in order to identify candidate biomarkers for sPTL using LC-MS/MS and reported eleven proteins uniquely expressed in sPTL complicated pregnancies [38].

CANDIDATE SERUM PROTEOMIC BIOMARKERS FOR SPONTANEOUS PRETERM LABOUR

At present, there have been comparatively limited reports focusing on identifying serum protein signatures related to sPTL following comparisons between serum samples obtained from women who subsequently experienced sPTL and uncomplicated term pregnancies. Nevertheless, maternal peripheral blood (serum or plasma), which is obtained non-invasively (or minimally invasively), represents a source of novel biomarkers that have the potential to advance prenatal screening for sPTL.

Esplin et al. were the first to report on serum proteins specific for sPTL at 24 and 28 weeks of gestation in asymptomatic women who subsequently developed sPTL. Using cLC-ESI-TOFMS they identified three specific peptides arising from the inter-alpha-trypsin inhibitor heavy chain 4 protein to be significantly under expressed in women at risk for sPTL. The most discriminating peptide demonstrated 65.0% sensitivity and 82.5% specificity of OR = 8.8, CI: 3.1 24.8. A combination of the three novel biomarkers and six previously studied biomarkers increased sensitivity up to 86.5% with a specificity of 80.6% at 28 weeks [39].

Parry et al. using pooled serum samples obtained at 19-24 weeks of gestation (n = 32) and 28 -32 weeks (n = 70) from women who had previously experienced sPTL found altered expression of 31 proteins in the sPTL group (< 34 weeks) as compared to term deliveries using liquid chromatography-multiple-reaction monitoring-mass spectrometry. Protein differential expression was further verified by Western blot in the placenta and fetal membranes. Notably, SERPIN B7, yielded a 1.5-fold increased mean concentration in women with subsequent preterm deliveries as compared to controls at 28 - 32 weeks whereas no difference was noted at 19 - 24 weeks. Furthermore, higher levels of SERPIN B7 at both gestational age windows were associated with a shorter interval to delivery and higher levels of SERPIN B7 in samples from 28 - 32 weeks were associated with a lower gestational age at delivery. These results, however, require validation in an extended cohort [40].

Saade et al., sought to identify differentially expressed plasma proteins in women who subsequently developed sPTL in a large-scale study including 5501 pregnant women recruited between 2011 and 2013 as part of the Proteomic Assessment of Preterm Risk study performed at 11 sites in the USA. The study revealed that two proteins namely insulin-like growth factor-binding protein 4 (IBP4) and sex hormone- binding globulin (SHBG) may be used as predictors of sPTL with a ROC curve value of 0.75 and sensitivity and specificity of 0.75 and 0.74, respectively. At this sensitivity and specificity, the IBP4/SHBG demonstrated an odds ratio of 5.04 for sPTL.

Importantly, higher-risk subjects defined by the IBP4/SHBG predictor score generally gave birth earlier than lower-risk subjects thus enabling the identification of asymptomatic women at risk of sPTL in clinical practice and enhancing preventive interventions for the management of the complication [16].

Cantowine et al. reported on biomarkers that may allow for the development of a non-invasive test for the early prediction of sPTL before clinical presentation. Serum samples for this retrospective study were obtained from pregnant women at 10-12 weeks of gestation. Samples from 25-singleton sPTL cases that delivered ≤ 34 weeks and 50 from uncomplicated term deliveries matched by maternal age, race and gestational age of sampling were included in the study. In total 132 candidate biomarkers, mainly linked to inflammation, wound healing and the coagulation cascade, were identified by multiple reaction monitoring mass spectrometry. Following statistical analysis, 62 demonstrated robust power for detecting sPTL at an estimated false discovery rate of < 20%. Linear modeling using a multiplex of the candidate biomarkers with a fixed sensitivity of 80% exhibited a specificity of 83% with a median area under the curve of 0.89. These results indicate a strong potential of multivariate model development for risk stratification [41].

Lynch et al. identified the complement factors B and H as well as the coagulation factors IX and IX ab as candidate biomarkers for sPTL at 10-15 weeks of gestation implying an association between immune and coagulation events in early pregnancy with subsequent sPTL [42].

In another study carried out aiming to detect proteomic predictors of premature delivery, Gunko *et al.* reported 13 serum proteins down-regulated and 12 up-regulated at 16-17 weeks of gestation in women at high risk for sPTL. Altered expressed proteins are known to be implicated in various regulatory and molecular functions including antioxidant enzymes, chaperones, cytoskeleton proteins, cell adhesion molecules, angiogenesis, proteolysis, transcription, and inflammation [43].

Finally in a more recent study D'Silva *et al.,* applied 2DE coupled with MS to analyze first trimester maternal serum. They reported 30 proteoforms significantly altered expressed in the sPTL group as compared to uncomplicated pregnancies including nine phosphoproteins and eleven glycoproteins.

Changes occurred in proteins linked to immune and defense responses implying a role of proteoforms in maternal serum as mediators of the disorder [5].

CONCLUSION

Up to now, a broad spectrum of candidate biomarkers for sPTL, mainly involved in inflammation, coagulation, and proteolysis has been identified using proteomic technology. Nevertheless, inconsistencies exist among studies, and no similarities have been observed which can be attributed to variations in methodologies in the pre-analytical and analytical level. Therefore, future studies in this field should involve the application of more sensitive methods followed by validation of the biomarkers in large-scale trials. Since candidate biomarkers are low abundant proteins, methods to deplete the high abundant proteins should also be considered to improve the quality of protein separation and the number of identified proteins.

Most importantly, in order to advance screening, novel biomarkers should be detected in biological fluids obtained non-invasively such as maternal blood.

Ideally, biomarkers for sPTL should predict the complication early during pregnancy, in the late 1st trimester (11 - 13 weeks gestation) when screening for fetal aneuploidies is also performed. Consequently, a predictive screening test for sPTL may be incorporated to prenatal screening routinely offered to all pregnant women in the late 1st trimester to provide early reassurance of the parents for the wellbeing of the fetus and pregnancy outcome.

REFERENCES

[1] Suff, N., L. Story, A. Shennan, The prediction of preterm delivery: What is new? *Semin. Fetal Neonatal Med.,* 24(1) (2019) 27 - 32.

[2] Moutquin, J. M., Classification and heterogeneity of preterm birth, *BJOG,* 110 Suppl. 20 (2003) 30 - 3.

[3] Frey, H. A., M. A. Klebanoff. The epidemiology, etiology, and costs of preterm birth, Semin. *Fetal Neonatal Med.,* 21(2) (2016) 68 - 73.

[4] Lotfi, G., S. Faraz, R. Nasir, S. Somini, R. M. Abdeldayem, R. Koratkar, N. Alsawalhi, A. Ammar. Comparison of the effectiveness of a PAMG-1 test and standard clinical assessment in the prediction of preterm birth and reduction of unnecessary hospital admissions, *J. Matern. Fetal Neonatal Med.*, (2017) 1 - 5.

[5] D'Silva, A. M., J. A. Hyett, J. R. Coorssen. Proteomic analysis of first trimester maternal serum to identify candidate biomarkers potentially predictive of spontaneous preterm birth, *J. Proteomics,* 178 (2018) 31 - 42.

[6] G. B. D. DALYs, H. Collaborators, Global, regional, and national disability-adjusted life-years (DALYs) for 315 diseases and injuries and healthy life expectancy (HALE), 1990-2015: a systematic analysis for the Global Burden of Disease Study 2015, *Lancet,* 388(10053) (2016) 1603 - 1658.

[7] Lucovnik, M., A. T. Bregar, L. Steblovnik, I. Verdenik, K. Gersak, I. Blickstein, N. Tul, Changes in incidence of iatrogenic and spontaneous preterm births over time: a population-based study, *J. Perinat. Med.,* 44(5) (2016) 505 - 9.

[8] Alfirevic, Z., T. Stampalija, N. Medley, Cervical stitch (cerclage) for preventing preterm birth in singleton pregnancy, *Cochrane Database Syst. Rev*., 6 (2017) CD008991.

[9] Hezelgrave, N. L., K. Kuhrt, K. Cottam, P. T. Seed, R. M. Tribe, A. H. Shennan, The effect of blood staining on cervicovaginal quantitative fetal fibronectin concentration and prediction of spontaneous preterm birth, *Eur. J. Obstet. Gynecol. Reprod. Biol.,* 208 (2017) 103 - 108.

[10] Jarde, A., O. Lutsiv, C. K. Park, J. Beyene, J. M. Dodd, J. Barrett, P. S. Shah, J. L. Cook, S. Saito, A. B. Biringer, L. Sabatino, L. Giglia, Z. Han, K. Staub, W. Mundle, J. Chamberlain, S. D. McDonald, Effectiveness of progesterone, cerclage and pessary for preventing preterm birth in singleton pregnancies: a systematic review and network meta-analysis, *BJOG,* 124(8) (2017) 1176 - 1189.

[11] Romero, R., K. H. Nicolaides, A. Conde-Agudelo, J. M. O'Brien, E. Cetingoz, E. Da Fonseca, G. W. Creasy, S. S. Hassan, Vaginal

progesterone decreases preterm birth </= 34 weeks of gestation in women with a singleton pregnancy and a short cervix: an updated meta-analysis including data from the OPPTIMUM study, *Ultrasound Obstet. Gynecol.,* 48(3) (2016) 308 - 17.

[12] Mackeen, A. D., T. J. Rafael, J. Zavodnick, V. Berghella, Effectiveness of 17-alpha-hydroxyprogesterone caproate on preterm birth prevention in women with history-indicated cerclage, *Am. J. Perinatol.,* 30(9) (2013) 755 - 8.

[13] Georgiou, H. M., M. K. Di Quinzio, M. Permezel, S. P. Brennecke, Predicting Preterm Labour: Current Status and Future Prospects, *Dis. Markers,* 2015 (2015) 435014.

[14] Laughon, S. K., P. S. Albert, K. Leishear, P. Mendola, The NICHD Consecutive Pregnancies Study: recurrent preterm delivery by subtype, *Am. J. Obstet. Gynecol.,* 210(2) (2014) 131 e1-8.

[15] Slattery, M. M., J. J. Morrison, Preterm delivery, *Lancet,* 360(9344) (2002) 1489 - 97.

[16] Saade, G. R., K. A. Boggess, S. A. Sullivan, G. R. Markenson, J. D. Iams, D. V. Coonrod, L. M. Pereira, M. S. Esplin, L. M. Cousins, G. K. Lam, M. K. Hoffman, R. D. Severinsen, T. Pugmire, J. S. Flick, A. C. Fox, A. J. Lueth, S. R. Rust, E. Mazzola, C. Hsu, M. T. Dufford, C. L. Bradford, I. E. Ichetovkin, T. C. Fleischer, A. D. Polpitiya, G. C. Critchfield, P. E. Kearney, J. J. Boniface, D. E. Hickok, Development and validation of a spontaneous preterm delivery predictor in asymptomatic women, *Am. J. Obstet. Gynecol.,* 214(5) (2016) 633 e1-633 e24.

[17] La Merrill, M., C. R. Stein, P. Landrigan, S. M. Engel, D. A. Savitz, Prepregnancy body mass index, smoking during pregnancy, and infant birth weight, *Ann. Epidemiol.,* 21(6) (2011) 413 - 20.

[18] Gundu, S., M. Kulkarni, S. Gupte, A. Gupte, M. Gambhir, P. Gambhir, Correlation of first-trimester serum levels of pregnancy-associated plasma protein A with small-for-gestational-age neonates and preterm births, *Int. J. Gynaecol. Obstet.,* 133(2) (2016) 159 - 63.

[19] Krispin, E., E. Hadar, R. Chen, A. Wiznitzer, B. Kaplan, The association of different progesterone preparations with preterm birth prevention, *J. Matern. Fetal Neonatal Med.*, (2018) 1 - 6.

[20] Beta, J., R. Akolekar, W. Ventura, A. Syngelaki, K. H. Nicolaides, Prediction of spontaneous preterm delivery from maternal factors, obstetric history and placental perfusion and function at 11-13 weeks, *Prenat. Diagn.,* 31(1) (2011) 75 - 83.

[21] Greco, E., R. Gupta, A. Syngelaki, L. C. Poon, K. H. Nicolaides, First-trimester screening for spontaneous preterm delivery with maternal characteristics and cervical length, *Fetal Diagn. Ther.,* 31(3) (2012) 154 - 61.

[22] Mella, M. T., A. D. Mackeen, D. Gache, J. K. Baxter, V. Berghella, The utility of screening for historical risk factors for preterm birth in women with known second trimester cervical length, *J. Matern. Fetal Neonatal Med.*, 26(7) (2013) 710 - 5.

[23] Law, K. P., T. L. Han, C. Tong, P. N. Baker, Mass spectrometry-based proteomics for pre-eclampsia and preterm birth, *Int. J. Mol. Sci.*, 16(5) (2015) 10952 - 85.

[24] Wright, P. C., J. Noirel, S. Y. Ow, A. Fazeli, A review of current proteomics technologies with a survey on their widespread use in reproductive biology investigations, *Theriogenology,* 77(4) (2012) 738 - 765 e52.

[25] Anagnostopoulos, A. K., A. Kolialexi, A. Mavrou, K. Vougas, N. Papantoniou, A. Antsaklis, E. Kanavakis, M. Fountoulakis, G. T. Tsangaris, Proteomic analysis of amniotic fluid in pregnancies with Klinefelter syndrome foetuses, *J. Proteomics*, 73(5) (2010) 943 - 50.

[26] Anagnostopoulos, A. K., G. T. Tsangaris, Proteomics advancements in fetomaternal medicine, *Clin. Biochem.*, 46(6) (2013) 487 - 96.

[27] Mertins, P., N. D. Udeshi, K. R. Clauser, D. R. Mani, J. Patel, S. E. Ong, J. D. Jaffe, S. A. Carr, iTRAQ labeling is superior to mTRAQ for quantitative global proteomics and phosphoproteomics, *Mol. Cell. Proteomics,* 11(6) (2012) M111 014423.

[28] Unwin, R. D. Quantification of proteins by iTRAQ, *Methods Mol. Biol.,* 658 (2010) 205 - 15.

[29] Zhang, Y., B. R. Fonslow, B. Shan, M. C. Baek, J. R. Yates, 3rd, Protein analysis by shotgun/bottom-up proteomics, *Chem. Rev.,* 113 (4) (2013) 2343 - 94.

[30] Heng, Y. J., S. Liong, M. Permezel, G. E. Rice, M. K. Di Quinzio, H. M. Georgiou, Human cervicovaginal fluid biomarkers to predict term and preterm labor, *Front. Physiol.,* 6 (2015) 151.

[31] Liong, S., M. K. Di Quinzio, Y. J. Heng, G. Fleming, M. Permezel, G. E. Rice, H. M. Georgiou, Proteomic analysis of human cervicovaginal fluid collected before preterm premature rupture of the fetal membranes, *Reproduction,* 145(2) (2013) 137 - 47.

[32] Yuan, W., L. Chen, A. L. Bernal, Is elevated maternal serum alpha-fetoprotein in the second trimester of pregnancy associated with increased preterm birth risk? A systematic review and meta-analysis, *Eur. J. Obstet. Gynecol. Reprod. Biol.,* 145(1) (2009) 57 - 64.

[33] Romero, R., J. P. Kusanovic, F. Gotsch, O. Erez, E. Vaisbuch, S. Mazaki-Tovi, A. Moser, S. Tam, J. Leszyk, S. R. Master, P. Juhasz, P. Pacora, G. Ogge, R. Gomez, B. H. Yoon, L. Yeo, S. S. Hassan, W. T. Rogers, Isobaric labeling and tandem mass spectrometry: a novel approach for profiling and quantifying proteins differentially expressed in amniotic fluid in preterm labor with and without intra-amniotic infection/inflammation, *J. Matern. Fetal Neonatal Med.,* 23(4) (2010) 261 - 80.

[34] Shah, S. J., K. H. Yu, V. Sangar, S. I. Parry, I. A. Blair, Identification and quantification of preterm birth biomarkers in human cervicovaginal fluid by liquid chromatography/tandem mass spectrometry, *J. Proteome. Res.,* 8(5) (2009) 2407 - 17.

[35] Consonni, S., V. Mainini, A. Pizzardi, E. Gianazza, C. Chinello, A. Locatelli, F. Magni, Non-invasively collected amniotic fluid as a source of possible biomarkers for premature rupture of membranes investigated by proteomic approach, *Arch. Gynecol. Obstet.,* 289(2) (2014) 299 - 306.

[36] Pan, H. T., H. G. Ding, M. Fang, B. Yu, Y. Cheng, Y. J. Tan, Q. Q. Fu, B. Lu, H. G. Cai, X. Jin, X. Q. Xia, T. Zhang, Proteomics and bioinformatics analysis of altered protein expression in the placental villous tissue from early recurrent miscarriage patients, *Placenta,* 61 (2018) 1 - 10.

[37] Pereira, L., A. P. Reddy, T. Jacob, A. Thomas, K. A. Schneider, S. Dasari, J. A. Lapidus, X. Lu, M. Rodland, C. T. Roberts, Jr., M. G. Gravett, S. R. Nagalla, Identification of novel protein biomarkers of preterm birth in human cervical-vaginal fluid, *J. Proteome. Res.,* 6(4) (2007) 1269 - 76.

[38] Butt, R. H., M. W. Lee, S. A. Pirshahid, P. S. Backlund, S. Wood, J. R. Coorssen, An initial proteomic analysis of human preterm labor: placental membranes, *J. Proteome. Res.,* 5(11) (2006) 3161 - 72.

[39] Esplin, M. S., K. Merrell, R. Goldenberg, Y. Lai, J. D. Iams, B. Mercer, C. Y. Spong, M. Miodovnik, H. N. Simhan, P. van Dorsten, M. Dombrowski, H. Eunice Kennedy Shriver National Institute of Child, N. Human Development Maternal-Fetal Medicine Units, Proteomic identification of serum peptides predicting subsequent spontaneous preterm birth, *Am. J. Obstet. Gynecol.,* 204(5) (2011) 391 e1-8.

[40] Parry, S., H. Zhang, J. Biggio, R. Bukowski, M. Varner, Y. Xu, W. W. Andrews, G. R. Saade, M. S. Esplin, R. Leite, J. Ilekis, U. M. Reddy, Y. Sadovsky, I. A. Blair, H. Eunice Kennedy Shriver National Institute of Child, G. Human Development, R. Proteomic Network for Preterm Birth, Maternal serum serpin B7 is associated with early spontaneous preterm birth, *Am. J. Obstet. Gynecol.,* 211(6) (2014) 678 e1-12.

[41] Cantonwine, D. E., Z. Zhang, K. Rosenblatt, K. S. Goudy, R. C. Doss, A. M. Ezrin, G. Page, B. Brohman, T. F. McElrath, Evaluation of proteomic biomarkers associated with circulating microparticles as an effective means to stratify the risk of spontaneous preterm birth, *Am. J. Obstet. Gynecol.,* 214(5) (2016) 631 e1-631 e11.

[42] Lynch, A. M., B. D. Wagner, R. R. Deterding, P. C. Giclas, R. S. Gibbs, E. N. Janoff, V. M. Holers, N. F. Santoro, The relationship of circulating proteins in early pregnancy with preterm birth, *Am. J. Obstet. Gynecol.,* 214(4) (2016) 517 e1-517 e8.

[43] Gunko, V. O., T. N. Pogorelova, V. A. Linde, Proteomic Profiling of the Blood Serum for Prediction of Premature Delivery, *Bull. Exp. Biol. Med.,* 161(6) (2016) 829 - 832.

In: Proteomics
Editor: Ricardo Parker
ISBN: 978-1-53616-440-4

Chapter 4

PROTEOME ANALYSIS OF SOFT-BIOMASS MICROBIAL-DEGRADATION USING MONOLITH-BASED NANO-LC/MS/MS

Shunsuke Aburaya, Wataru Aoki and Mitsuyoshi Ueda*
Laboratory of Biomacromolecular Chemistry,
Department of Applied Biochemistry,
Division of Applied Life Sciences,
Graduate School of Agriculture, Kyoto University,
Kitashirakawa-Oiwake-cho, Sakyo-ku, Kyoto, Japan

ABSTRACT

Lignocellulosic biomass has gained much attention as an alternative and renewable carbon source. Lignocellulosic biomass has a highly complex and robust structure, being made up of cellulose, hemicellulose, pectin and lignin. Therefore, the degradation of lignocellulosic biomass to simple sugars is difficult due to its structure. For degradation of such lignocellulosic biomass, an anaerobic bacterium *Clostridium cellulovorans* has attractive features. *C. cellulovorans* secretes a large enzyme complex

* Corresponding Author's E-mail: miueda@kais.kyoto-u.ac.jp.

called "cellulosome" and this complex can degrade lignocellulosic biomass efficiently. For more efficient degradation of lignocellulosic biomass, *C. cellulovorans* optimizes profiles of cellulosomal, and non-cellulosomal enzymes depending on carbon sources. In order to understand the mechanisms of efficient degradation of lignocellulosic biomass, comprehensive analysis of carbohydrate-degrading and metabolic enzymes is required. In this chapter, we summarize proteome analysis of *C. cellulovorans* cultured with different carbon sources, and provide some insights how *C. cellulovorans* optimizes diverse enzymes depending on carbon sources.

Keywords: *Closridium cellulovorans*, proteome analysis, monolithic column

INTRODUCTION

Humanity's lifestyle relies on fossil fuels, but the overconsumption of such fuels gives rise to too much CO_2 emission and environmental pollution (Mood et al., 2013). Furthermore, the supply of fossil fuels is limited (Mood et al., 2013). Therefore, new resources that can act as alternatives to fossil fuels are required. Such alternative resources have some requirements; that is, they should be renewable and carbon neutral to overcome the limitations in fossil fuels, and they should be able to manufacture products that are usually made from fossil fuels. Plant biomass has attracted much attention as an alternative resource to satisfy these requirements, and can be used to produce bioethanol and useful chemical products.

PROBLEMS WITH PLANT BIOMASS

Plant biomass has the potential to overcome the limitation of fossil fuels because it is a renewable and carbon neutral resource (Lee, 1997; Menon and Rao, 2012; Mood et al., 2013). Nowadays, in Brazil, bioethanol production is being conducted using the starch from edible polysaccharides, such as corn and sugarcane. However, bioethanol production from edible polysaccharides remains problematic. First, expanding edible biomass production for fuel creates a poor energy balance (energy output from the biofuel/energy inputs used for production) and has negative impacts on the regional water resources, biodiversity, and soil quality (Groom et al., 2008). Second, a vast amount of agrowastes is released from edible biomass-based biofuel production (Bjerre et al., 1996). Third, the use of limited agricultural land for producing biomass for fuel competes with food production. Therefore, new biomass sources are required to overcome these dilemmas (Sun and Cheng, 2002).

Currently, lignocellulosic biomass has gained much attention as an alternative and renewable resource (Lynd et al., 2005). This biomass type is a cheap, abundant, and renewable carbon source, being obtained as agricultural byproducts and industrial garbage (Lynd et al., 1999).

Table 1. Composition of the polysaccharides in major lignocellulosic biomass materials

Lignocellulosic biomass	Cellulose	Hemicellulose				Others	Ref.
	Glucan	Xylan	Arabinan	Galactan	Mannan		
Corn cob	34.3	31.1	3.0	-	-	31.6	(Garrote et al., 2007)
Corn stover	38.2	21.0	2.7	2.1	-	36.0	(Li et al., 2010c)
Rice straw	31.1	18.7	3.6	-	-	46.6	(Chen et al., 2011)
Switchgrass	39.5	20.3	2.1	2.6	-	35.5	(Li et al., 2010a)
Sugarcane bagasse	43.3	22.4	2.3	0.6	0.4	31.0	(Girisuta et al., 2013)
Wheat straw	39.8	34.5	2.8	-	-	22.9	(Kristensen et al., 2008)

All values are percentages [%] based on dry weight.

Lignocellulosic biomass has a highly complex and robust structure, being made up of cellulose, hemicellulose, pectin, and lignin, making its hydrolysis to simple sugars difficult (Agbor et al., 2011; Mood et al., 2013). The major polysaccharide components of some lignocellulosic biomass materials are shown in Table 1. Although lignocellulosic biomass is composed mainly of cellulose, some materials contain mostly hemicellulose.

Therefore, all polysaccharide components of lignocellulosic biomass have to be degraded.

STRUCTURAL POLYSACCHARIDES OF THE PLANT CELL WALL

Cellulose

Cellulose, a major component of the plant cell wall, is the most abundant polysaccharide in lignocellulosic biomass. It is a linear and unbranched homopolymer of ß-1,4-glycosidic bonds. Each cellulose chain is bundled together to construct cellulose microfibrils that are linked by intramolecular or intermolecular hydrogen bonds (Li et al., 2010b) and van der Waals forces (Agbor et al., 2011). This structure makes cellulose difficult to degrade with enzymes.

Hemicellulose

Hemicellulose has a complex structure made up of pentose (xylose and arabinose), hexose (mannose, glucose, and galactose), and acetylated sugars. The main component of hemicellulose in hard wood, straw, and grass is xylan, whereas it is galactomannan in soft wood (Agbor et al., 2011). Hemicellulose is thought to crosslink cellulose and lignin, with the crosslinked structure forming the entire structure of the plant cell wall (Cosgrove, 2005; Hayashi, 1989). Hemicellulose has a branched and amorphous structure and a low molecular weight, and because it is sensitive to enzymes, it degrades more easily than cellulose does (Li et al., 2010b; Saha, 2003). Therefore, the degradation of hemicellulose increases the degradation efficiency of lignocellulosic biomass (Agbor et al., 2011).

Pectin

Pectin is a family of galacturonic acid-rich polysaccharides composed of homogalacturonan, rhamnogalacturonan I, and rhamnogalacturonan II (Mohnen, 2008). Pectin has the most complex polysaccharide structure in the living world (Willats et al., 2001), and these polysaccharides are thought to be crosslinked together via covalent bonds (Cumming et al., 2005; Vincken et al., 2003). Furthermore, it was suggested that pectins may serve to hold some hemicelluloses in the cell wall (Cumming et al., 2005).

Homogalacturonan is a linear homopolymer of galacturonic acid, and is methylesterified at the C-6 carboxyl residues and *O*-acetylated at O-2 and O-3 (O'Neill et al., 1990).

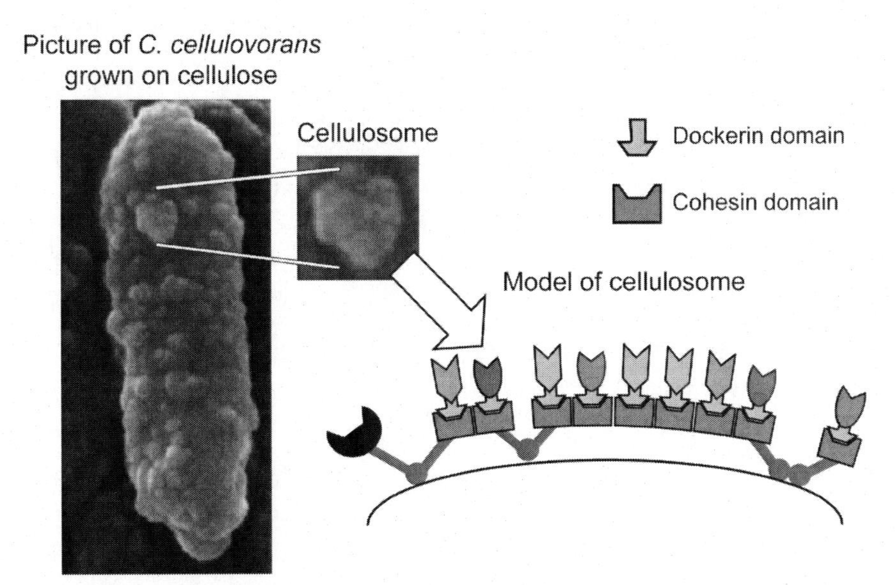

(Modified from (Blair and Anderson, 1999)).

Figure 1. Schematic diagram of the cellulosome of *Clostridium cellulovorans*.

Rhamnogalacturonan I is composed of galacturonic acid, rhamnose, α-L-arabinofuranosyl, and β-D-galactopyranosyl residues (Mohnen, 2008). The type and number of sugars are quite different according to the developmental stages and cell types (Ridley et al., 2001).

Rhamnogalacturonan II has 12 different sugar residues in its main structure (Mohnen, 2008), and it forms a dimer with the borate ester (Kobayashi et al., 1996).

CELLULOSOME

Cellulosomes, which are localized at the bacterial cell surface, are valuable multienzymatic complexes composed of scaffoldins and enzymes for the degradation of lignocellulosic biomass (Figure 1) (Artzi et al., 2017; Shoham et al., 1999). The scaffoldins play a role in the assembly of other cellulosomal proteins and in binding to carbohydrates (Artzi et al., 2017; Shoham et al., 1999). The scaffoldins are made up of cohesin domains that interact with a dockerin domain, and are the carbohydrate-binding modules that bind to and arrange enzymes at appropriate sites on the plant cell wall (Artzi et al., 2017; Shoham et al., 1999). The dockerin domain is usually fused with carbohydrate-active enzymes (cellulosomal proteins) (Artzi et al., 2017; Shoham et al., 1999). In the cellulosome, these carbohydrate-active enzymes with dockerin domains are proximally positioned by the consecutive cohesin domains (Artzi et al., 2017; Shoham et al., 1999). Therefore the assembled carbohydrate-active enzymes react synergistically, and cellulosomes have a higher activity than free carbohydrate-active enzymes do for the degradation of polysaccharides (Murashima et al., 2003).

With regard to their structure, there are two types of cellulosomes: simple cellulosomes and highly structured cellulosomes. Simple cellulosomes are produced by mesophilic bacteria, such as *Clostridium cellulolyticum* (Gal et al., 1997), *Clostridium cellulovorans* (Doi and Tamaru, 2001), *Ruminococcus albus* (Ohara et al., 2000), and *Ruminococcus bromii* (Cann et al., 2016). A simple cellulosome is constructed from a single primary scaffoldin and does not have a complex structure (Figure 2 (a)) (Artzi et al., 2017). Highly structured cellulosomes are produced by thermophilic and mesophilic bacteria, such as *Clostridium thermocellum* (Bayer et al., 1985) and *Acetivibrio cellulolyticus* (Xu et al., 2003). A highly structured cellulosome is constructed from multiple scaffoldins, and a single complex has many scaffoldins and carbohydrate-degrading enzymes (Bayer et al., 1998). The mechanism by which simple cellulosomes bind to the cell surface has not been elucidated. On the other hand, the mechanism behind the binding of highly structured cellulosomes to the cell surface has been suggested (Lemaire et al., 1995). First, one of the dockerins in the primary scaffoldin binds with cohesin at the cell surface. This cohesin molecule then connects either with the peptidoglycan-binding S-layer homology domain by noncovalent binding, or with the sortase motif by covalent binding (Lemaire et al., 1995).

CELL-FREE CELLULOSOMES

It was revealed that some cellulosomes were not localized at the cell surface recently. Some highly structured cellulosomes produced by *C. thermocellum* and *Clostridium clariflavum* do not localize at the cell surface, but act as secreted protein complexes instead (Artzi et al., 2015; Xu et al., 2016). These cellulosomes form a complex with cellulosomal proteins without binding to the cell surface, and are known as "cell-free cellulosomes" (Figure 2 (b)). In *C. cellulolyticum*, the enzymatic activities from cell-free cellulosomes have been determined (Mohand-Oussaid et al., 1999). Furthermore, it is known that cellulosomes at the cell surface detach from the surface at the late stationary phase in *C. thermocellum* (Bayer and Lamed, 1986).

Clostridium cellulovorans

C. cellulovorans, which was isolated from fermented wood biomass in 1984 (Sleat et al., 1984), is a mesophilic and anaerobic gram-positive bacterium (Figure 3). Recently, *C. cellulovorans* has garnered much attention as a cellulosome producer (Blair and Anderson, 1999; Sleat et al., 1984).

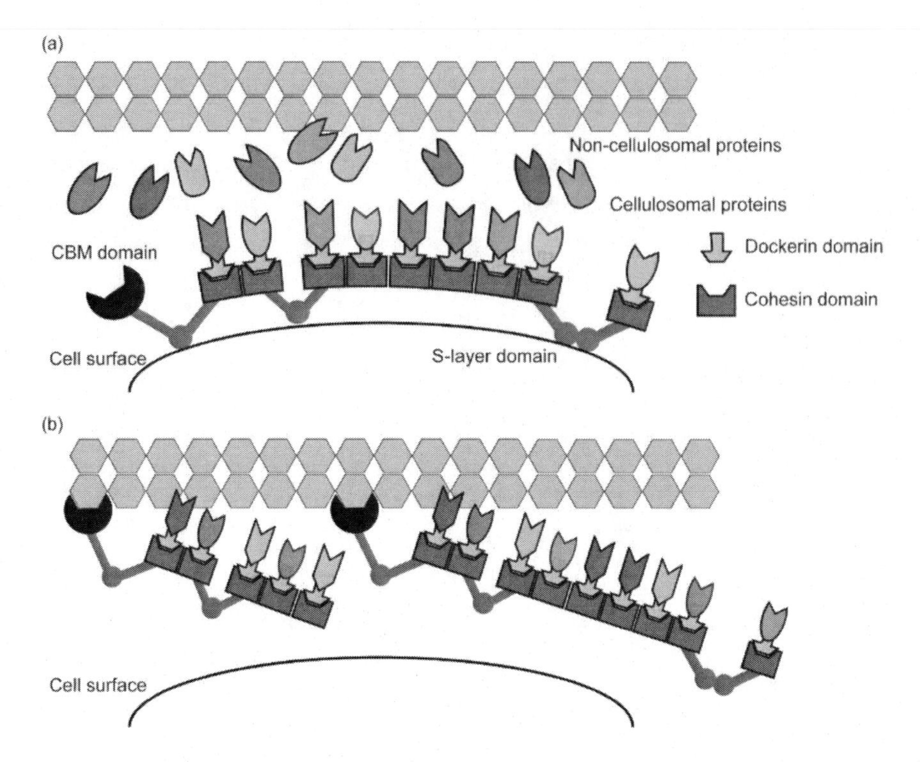

Figure 2. (a) Model of a simple cellulosome. (b) Model of a cell-free cellulosome. CBM: carbohydrate-binding module.

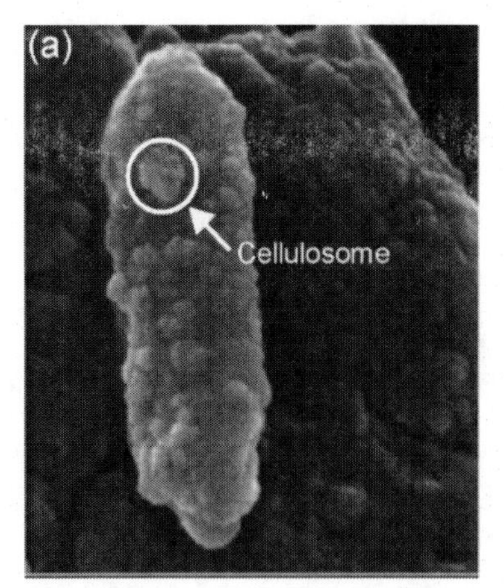

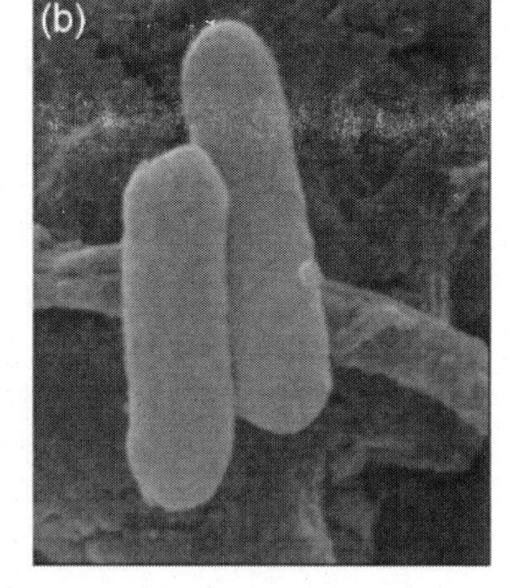

(Modified from (Blair and Anderson, 1999)).

Figure 3. (a) Cell surface structure of *Clostridium cellulovorans* grown on cellulose (b) Cell surface structure of *C. cellulovorans* grown on cellobiose.

It can degrade and metabolize the major plant cell-wall polysaccharides of lignocellulosic biomass, such as cellulose, hemicellulose, and pectin (Sleat et al., 1984). *C. cellulovorans* produces simple cellulosomes, the scaffoldin of which is constructed from nine cohesin domains (Blair and Anderson, 1999; Doi et al., 1994). Furthermore, the bacterium degrades plant cell-wall polysaccharides with both cellulosomal and non-cellulosomal proteins (Doi et al., 1998; Morisaka et al., 2012), the latter of which are secreted carbohydrate-degrading proteins without a dockerin domain (Doi et al., 1998).

Genome Analysis of *C. Cellulovorans*

In 2010, Tamaru et al. clarified the genome sequence of *C. cellulovorans*, offering new insight into the bacterium's efficient mechanism of lignocellulosic biomass degradation, (Tamaru et al., 2011; Tamaru et al., 2010). *C. cellulovorans* has 57 cellulosomal genes in its genome, which include those codings for 25 glycoside hydrolase (GH) family proteins, 2 carbohydrate esterase (CE) family proteins, and 4 polysaccharide lyase (PL) family proteins. In addition, the bacterium has 93 non-cellulosomal proteins, which comprise 53 GH family proteins, 5 CE family proteins, 10 PL family proteins, and 25 glycosyltransferase (GT) family proteins. In *C. cellulovorans*, the ratio of cellulosomal GH and PL family proteins to non-cellulosomal GH and PL family proteins is lower than that of other cellulosome-producing *Clostridium* species, whereas the total number of GH and PL family proteins is higher (Table 2) (Tamaru et al., 2011; Tamaru et al., 2010). This implies that *C. cellulovorans* can degrade more types of plant cell-wall polysaccharides than other *Clostridium* species can, and emphasizes the importance of non-cellulosomal proteins for the degradation of plant cell-wall polysaccharides in this species. To reveal the mechanisms by which the cellulosomal and non-cellulosomal proteins degrade polysaccharides, we should analyze the changes of these proteins between different polysaccharides and time points.

Table 2. Number of cellulosomal and non-cellulosomal genes in cellulosome-producing *Clostridium* species (Tamaru et al., 2011)

Organism	Cellulosomal GHs + PLs	Non-cellulosomal GHs + PLs	Cellulosomal GHs + PLs/ Non-cellulosomal GHs + PLs
C. cellulovorans	29	63	0.46
C. cellulolyticum	47	42	1.1
C. thermocellum	53	14	3.8

GHs: Glycoside hydrolase family proteins; PLs: Polysaccharide lyase family proteins.

MONOLITHIC COLUMN

High sensitivity and highly efficient separation are required for performing high-performance proteome analysis with liquid chromatography (LC) - electrospray ionization (ESI) - tandem mass spectrometry (MS/MS). Proteome samples have a high complexity and a broad dynamic range (Michalski et al., 2011). Therefore, the good separation and detection of a complex peptide mixture are important for proteome analysis. Recently, the development of mass spectrometry, as represented by Orbitrap mass spectrometry, has increased the number of peptides identified per minute by about tenfold (Eliuk and Makarov, 2015; Hebert et al., 2014). However, the separation power in the LC column has not increased in like. The high separation power in a particle-packed LC column is achieved by minimizing the diameter of the packed particle, but doing so increases the column back pressure (Giddings, 1991). Therefore, high separation efficiency in a particle-packed column is technically difficult to achieve beyond the upper-pressure limit of the LC instrument, necessitating new separation material for the LC column. To solve this problem, the monolithic column was developed as a new separation medium (Tanaka et al., 2002), enabling the preparation of a meter-long capillary column from which highly efficient separation and sensitivity were obtained for proteome analysis (Iwasaki et al., 2010; Morisaka et al., 2012).

OPTIMIZATION OF LC-ESI-MS/MS IN PROTEOME ANALYSIS

Optimization of the instruments is required for proteome analysis with a higher sensitivity. Although recent developments in instruments have enabled high sensitivity and highly efficient separation, we could still not perform highly sensitive proteome analysis without optimization of the instruments. Optimization has crucial effects; for example, in LC, the appropriate selection of a column and gradient method is necessary to maximize the number of peptides eluted at a single time (Vollmer et al., 2004). In ESI, an appropriate spray voltage is required for a high ionization efficiency and low background. In MS, several parameters should be appropriately set for identifying the eluting peptides (Vollmer et al., 2004). Similar to these examples, the parameters of LC, ESI, and MS affect the number of proteins identified and the peak area of MS1 (Andrews et al., 2011; Kalli and Hess, 2012). Hence, we attempted to optimize such parameters to identify larger numbers of proteins in non-targeted proteome analysis.

The design of experiments (DoE) could be a powerful tool for optimizing the combination of LC, ESI, and MS parameters. In several studies, the optimization of LC or MS has been performed by changing the parameters one by one (Kalli and Hess, 2012; Xu et al., 2009). In fact, although these studies optimized LC or MS parameters, such optimization procedure could not detect two-factor interactions, which can be important for determining an optimal parameter set (Kalli and Hess, 2012; Xu et al., 2009). To solve this problem, a recent study used DoE for optimizing MS parameters for non-targeted proteome analysis (Andrews et al., 2011). That study adopted a fractional factorial design, which could detect two-factor interactions from approximately three parameters. Using the fractional factorial design, MS parameters were systematically optimized and the number of proteins identified was increased. However, LC parameters and two-factor interactions between LC and MS were not analyzed. For optimizing LC, ESI, and MS parameters simultaneously, other statistical tools are required. One such new statistical tool, "definitive screening design," was recently developed (Jones and Nachtsheim, 2011). This tool might enable the identification of main factors and two-factor interactions from several parameters with less experiments.

Recently, Aburaya et al. applied "definitive screening design" to optimize LC-ESI-MS/MS parameters to increase the number of identified proteins and average peak area of MS1 in proteomics (Aburaya et al., 2017). Furthermore, this analysis confirmed two-factor interactions between LC parameter and MS parameter (Dynamic exclusion – Gradient, Resolution – Analysis time). Then, the number of identified proteins creased 8.1% in this analysis. This study showed the usefulness of DoE in optimization of the instruments in proteomics.

PROTEOME ANALYSIS OF CELLULOSOMAL AND NON-CELLULOSOMAL PROTEINS

Proteome analysis is a valuable tool for analyzing the dynamics of cellulosomal and non-cellulosomal proteins. For analyzing the role of cellulosomal proteins in cellulosome-producing *Clostridium* species, the enzymatic activities of the cellulosomal proteins have been analyzed; for example, EngK (Arai et al., 2006), PelA (Tamaru and Doi, 2001), and XynA (Kosugi et al., 2002) in *C. cellulovorans*. However, these analyses could not clarify the difference between the distinct compositions of cellulosomal and non-cellulosomal proteins. For such elucidation, we require a comprehensive and less-biased proteome analysis, including that of non-cellulosomal proteins.

Previously, the analysis of cellulosomal proteins was mainly performed using two-dimensional polyacrylamide gel electrophoresis (2D-PAGE) (Blouzard et al., 2010; Cho et al., 2010). However, we could not detect a lot of proteins with 2D-PAGE because antibodies or cohesin markers were used for detecting cellulosomal proteins; hence, we could only identify cellulosomal proteins (Blouzard et al., 2010; Cho et al., 2010). Furthermore, previous studies had concentrated proteins with the carbohydrate-binding domain by phosphoric acid-swollen cellulose, and therefore proteome analysis of non-cellulosomal proteins could not be performed. The genome analysis of *C. cellulovorans* (Tamaru et al., 2010) enabled the non-targeted proteome analysis of the proteins secreted by this bacterium (Morisaka et al., 2012). Furthermore, ultrafiltration of the cell-free culture medium was applied for enrichment of the secreted proteins, thereby permitting us to perform a less-biased analysis of the cellulosomal and non-cellulosomal proteins (Morisaka et al., 2012). As a result, Matsui et al. showed that the variance of non-cellulosomal proteins has an important role in the degradation of xylan and pectin (Matsui et al., 2013). Furthermore, Esaka et al. performed exo-proteome analysis with natural soft-biomass degradation. This study also suggests the important role in non-cellulosomal proteins in degradation of natural soft-biomass (Esaka et al., 2015). Ever since this study, however, proteome analysis of the non-cellulosomal proteins of other cellulosome-producing bacteria has not been carried out.

PROTEOME ANALYSIS FOR METABOLISM OF CARBON SOURCES IN CELLULOSOME-PRODUCING *CLOSTRIDIUM* SPECIES

Cellulosome-producing *Clostridium* species can degrade most of the major plant cell-wall polysaccharides. However, each bacterium has different types of carbohydrate-degrading enzymes and metabolizes different carbon sources (Table 3). For example, *C. thermocellum* grows well on cellooligosaccharides, ranging from cellobiose to cellohexaose (Zhang and Lynd, 2005), but cannot metabolize xylan and other major plant cell-wall polysaccharides (Ng et al., 1977). *C. thermocellum* utilizes highly structured cellulosomes for degrading plant cell-wall polysaccharides, but this bacterium does not have many genes coding for non-cellulosomal proteins (Tamaru et al., 2011).

Table 3. Degradation of carbon sources or growth in carbon sources in cellulosome-producing *Clostridium* species

Clostridium species	Function	Cellulose	Xylan	Galactomannan	Pectin
C. cellulovorans	Degradation	o	o	o	o
	Growth	o	o	o	o
C. thermocellum	Degradation	o	o	o	o
	Growth	o	x	x	x
C. cellulolyticum	Degradation	o	o	o	o
	Growth	o	o	Δ	x

(o: Ability for full degradation of each polysaccharide or high growth; Δ: Ability for partial degradation or little growth; x: No ability for degradation or no growth).

C. cellulovorans shows a distinct system for degrading and metabolizing major plant cell-wall polysaccharides compared with the other cellulosome-producing *Clostridium* species. *C. cellulovorans* can metabolize cellulose, hemicellulose, and pectin (Sleat et al., 1984). Furthermore, it encodes 97 non-cellulosomal proteins in its genome, and degrades the plant cell wall efficiently through a combination of cellulosomal and non-cellulosomal proteins (Matsui et al., 2013; Tamaru et al., 2011).

In *C. thermocellum* and *C. cellulolyticum*, proteome analysis of the cellular proteins using LC-MS/MS and transcriptome analysis has been performed for analyzing the cellular dynamics (Rydzak et al., 2012; Xu et al., 2013). For example, *C. thermocellum* was cultured in cellobiose, and the differences in proteins for core metabolic functions between the different growth phases were clarified by proteome analysis using general LC-MS/MS of the cellular proteins (Rydzak et al., 2012). In *C. cellulolyticum*, transcriptome analysis revealed the sensing mechanism of carbon sources in the culture medium, and the regulatory component for expressing the proteins involved in carbohydrate degradation and metabolism under different carbon sources (Xu et al., 2013).

C. cellulovorans shows a distinct system for degrading and metabolizing major plant cell-wall polysaccharides compared with the other cellulosome-producing *Clostridium* species. *C. cellulovorans* can metabolize cellulose, hemicellulose, and pectin (Sleat et al., 1984).

Table 4. Substrate-specific proteins

Xylan specific proteins		Galactomannan specific proteins	
Locus	**Description**	**Locus**	**Description**
Clocel_0589	alpha-L-fucosidase	Clocel_0034	glycoside hydrolase family protein
Clocel_0590	xylose isomerase	Clocel_0391	glycosyltransferase
Clocel_0591	transaldolase	Clocel_0684	thiazole biosynthesis protein ThiH
Clocel_0592	xylulokinase	Clocel_2225	cobalamin biosynthesis CbiX protein
Clocel_1085	dinitrogenase iron-molybdenum cofactor biosynthesis protein	Clocel_2259	PfkB domain-containing protein
Clocel_1151	methyl-accepting chemotaxis sensory transducer	Clocel_2697	sialate O-acetylesterase
Clocel_1430	glycoside hydrolase family protein	Clocel_2800	Alpha-galactosidase
Clocel_2573	hypothetical protein Clocel_2573	Clocel_2962	inosine-5\'-monophosphate dehydrogenase
Clocel_2592	two component transcriptional regulator, AraC family	Clocel_3175	PhoH family protein
Clocel_2595	xylan 1,4-beta-xylosidase	Clocel_3194	mannose-6-phosphate isomerase
Clocel_2596	sugar ABC transporter periplasmic protein inner-membrane translocator	Clocel_3196	glycosidase-like protein
Clocel_2597	ABC transporter	Clocel_3198	N-acylglucosamine 2-epimerase
Clocel_2598	PTS system lactose/cellobiose-specific	Clocel_3200	binding-protein-dependent transport system inner membrane protein
Clocel_2881	transporter subunit IIB putative phosphate transport regulator	Clocel_3201	extracellular solute-binding protein
Clocel_2940	PhoH family protein	Clocel_3205	glycoside hydrolase family 2
Clocel_3175	ATP:guanido phosphotransferase	Clocel_3657	xylan 1,4-beta-xylosidase
Clocel_3761	UvrB/UvrC protein	Clocel_3857	ABC transporter
Clocel_3762	aldo/keto reductase	Clocel_4053	LPXTG-motif cell wall anchor domain-containing protein
Clocel_4277		Clocel_4087	aldose 1-epimerase
		Clocel_4088	galactose-1-phosphate uridylyltransferase
		Clocel_4089	UDP-glucose 4-epimerase
		Clocel_4277	aldo/keto reductase

Table 4. (Continued)

Pectin specific proteins			
Locus	**Description**	**Locus**	**Description**
Clocel_0048	transcriptional regulator, AbrB family	Clocel_2253	Crp family transcriptional regulator
Clocel_0322	TatD family hydrolase	Clocel_2254	glycoside hydrolase family protein
Clocel_0513	extracellular solute-binding protein	Clocel_2255	major facilitator superfamily protein
Clocel_0519	glycogen/starch/alpha-glucan phosphorylase	Clocel_2256	glycosyl hydrolase family protein
Clocel_1243	extracellular solute-binding protein	Clocel_2259	PfkB domain-containing protein
Clocel_1892	acetate kinase	Clocel_2262	short-chain dehydrogenase/reductase SDR
Clocel_2210	nicotinate-nucleotide--dimethylbenzimidazole phosphoribosyltransferase	Clocel_2263	4-deoxy-L-threo-5-hexosulose-uronate ketol-isomerase
Clocel_2214	ATP:corrinoid adenosyltransferase BtuR/CobO/CobP	Clocel_2403	glucosamine/fructose-6-phosphate aminotransferase
Clocel_2222	precorrin-3B C(17)-methyltransferase cobalamin (vitamin B12) biosynthesis	Clocel_2737	small GTP-binding protein
Clocel_2225	CbiX protein	Clocel_3380	LPXTG-motif cell wall anchor domain-containing protein
Clocel_2227	precorrin-4 C(11)-methyltransferase	Clocel_3909	quorum-sensing autoinducer 2 (AI-2), LuxS
Clocel_2250	altronate dehydratase	Clocel_4088	galactose-1-phosphate uridylyltransferase
Clocel_2251	mannitol dehydrogenase domain-containing protein	Clocel_4277	aldo/keto reductase

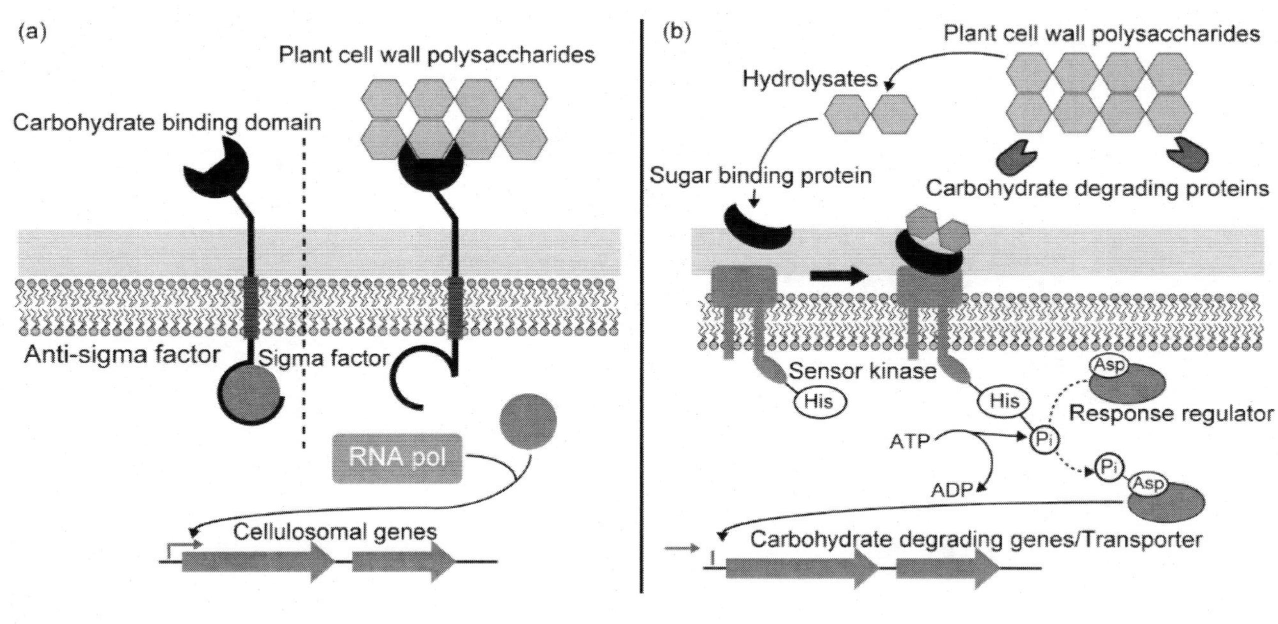

Figure 4. Plant cell-wall polysaccharide-sensing model in cellulosome-producing *Clostridium* species. (a) Plant cell-wall polysaccharide-sensing model in *C. thermocellum.* (b) Plant cell-wall polysaccharide-sensing model in *C. cellulolyticum.*

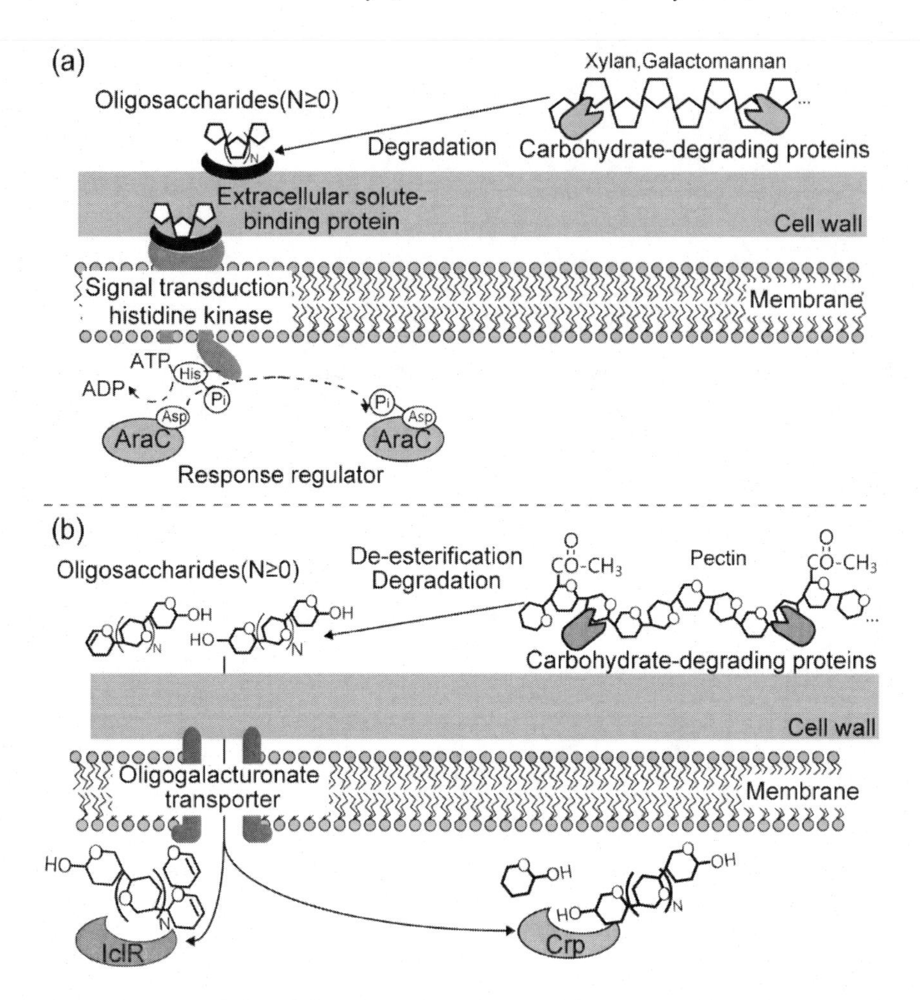

Figure 5. Proposed *C. cellulovorans* substrate recognition systems of hemicellulose and pectin. For hemicellulose, polysaccharides are degraded to derived oligosaccharides outside of the cell, and extracellular solute-binding proteins bind these substrates. Solute-binding proteins induce signal transduction integral membrane histidine kinases, and the activated kinases phosphorylate transcriptional regulator AraC, and target genes (Aburaya et al. 2015) are upregulated (a). For pectin, polysaccharides are de-esterified by pectinesterase and degraded by pectin lyase. Next, derived oligosaccharides are transported into cell, and these or other metabolites bind to a transcriptional regulator. Then, genes belonging to the target cluster (Aburaya et al. 2015) are upregulated (b).

Furthermore, it encodes 97 non-cellulosomal proteins in its genome, and degrades the plant cell wall efficiently through a combination of cellulosomal and non-cellulosomal proteins (Matsui et al., 2013; Tamaru et al., 2011).

In *C. cellulovorans*, proteome analysis of cellular proteins was performed for elucidating the difference between metabolic enzymes in various plant cell wall polysaccharides (Aburaya et al., 2015). In this study, glucose, xylan, galactomannan, and pectin were used as major plant cell wall polysaccharides. Then, metabolic pathway of each polysaccharides was up-produced in these major plant cell wall polysaccharides. In pectin, *C. cellulovorans* utilizes two distinct pectin metabolism pathways, called as hydrolase/isomerase and lyase/5-dehydro-4-deoxy-gluconate pathways (Table 4).

PROTEOME ANALYSIS FOR ELUCIDATING SENSING OF CARBON SOURCES AND REGULATION OF CARBOHYDRATE-DEGRADING ENZYMES IN CELLULOSOME-PRODUCING *CLOSTRIDIUM* SPECIES

Cellulosome-producing *Clostridium* species change their expression pattern of carbohydrate-degrading enzymes depending on the polysaccharides present in the culture medium. Therefore, some studies have attempted to reveal the mechanisms responsible for sensing the carbon sources in the culture media and for the changing expression pattern of carbohydrate-degrading enzymes. In *C. thermocellum*, sigma factor and anti-sigma factor are used for changing the expression pattern of carbohydrate-degrading enzymes (Figure 4 (a)) (Nataf et al., 2010). Sigma factors are released in the cell when polysaccharides in the culture medium bind to the carbohydrate-binding module domain. Then, the sigma factors interact with RNA polymerase and promote the transcription of carbohydrate-degrading enzymes. In *C. cellulolyticum*, a two-component regulatory system is used for sensing polysaccharides in the culture medium and regulating the carbohydrate-degrading enzymes (Figure 4 (b) (Xu et al., 2013)). In the two-component regulatory system of this bacterial species, hydrolysates of polysaccharides in the culture medium bind to sugar-binding proteins, thereby activating histidine kinase and expression of the carbohydrate-degrading enzymes.

In contrast to these cellulosome-producing bacteria, the mechanism for sensing polysaccharides in the culture medium and for changing the expression pattern of carbohydrate-degrading enzymes have not been clarified in *C. cellulovorans* until recently. Aburaya et al. reported comparative quantitative proteome analysis to elucidate the mechanisms for sensing polysaccharides in the culture medium (Aburaya et al., 2015). As a result of quantitative proteome analysis, it was suggested that *C. cellulovorans* utilizes two-component regulatory system for sensing hemicelluloses (xylan and galactomannan) and two different transcriptional factors for sensing pectin (Aburaya et al., 2015). Substrate recognition systems in *C. cellulovorans* are one of the survival strategies. In other bacteria, monosaccharides such as glucose and xylose are mainly transported, but this bacterium recognizes oligosaccharides of hemicellulose and pectin, and can optimize degradation-, metabolism-, and transport-related proteins' profiles according to the substrate faster than other bacteria. This feature may be suitable for competing with other anaerobic bacteria (Figure 5 (a) and (b)).

REFERENCES

Aburaya, S., Aoki, W., Minakuchi, H. & Ueda, M. (2017). Definitive screening design enables optimization of LC–ESI–MS/MS parameters in proteomics. *Biosci Biotech Bioch*, *81*, 2237-2243.

Aburaya, S., Esaka, K., Morisaka, H., Kuroda, K. & Ueda, M. (2015). Elucidation of the recognition mechanisms for hemicellulose and pectin in *Clostridium cellulovorans* using intracellular quantitative proteome analysis. *AMB Express*, *5*, 29.

Agbor, V. B., Cicek, N., Sparling, R., Berlin, A. & Levin, D. B. (2011). Biomass pretreatment: fundamentals toward application. *Biotechnol Adv*, *29*, 675-685.

Andrews, G. L., Dean, R. A., Hawkridge, A. M. & Muddiman, D. C. (2011). Improving proteome coverage on a LTQ-Orbitrap using design of experiments. *J Am Soc Mass Spectrom*, *22*, 773-783.

Arai, T., Kosugi, A., Chan, H., Koukiekolo, R., Yukawa, H., Inui, M. & Doi, R. H. (2006). Properties of cellulosomal family 9 cellulases from *Clostridium cellulovorans*. *Appl Microbiol Biotechnol*, *71*, 654-660.

Artzi, L., Bayer, E. A. & Moraïs, S. (2017). Cellulosomes: bacterial nanomachines for dismantling plant polysaccharides. *Nat Rev Microbiol*, *15*, 83.

Artzi, L., Morag, E., Barak, Y., Lamed, R. & Bayer, E. A. (2015). *Clostridium clariflavum*: key cellulosome players are revealed by proteomic analysis. *mBio*, *6*, e00411-00415.

Bayer, E., Setter, E. & Lamed, R. (1985). Organization and distribution of the cellulosome in *Clostridium thermocellum*. *J Bacteriol*, *163*, 552-559.

Bayer, E. A. & Lamed, R. (1986). Ultrastructure of the cell surface cellulosome of *Clostridium thermocellum* and its interaction with cellulose. *J Bacteriol*, *167*, 828-836.

Bayer, E. A., Shimon, L. J. W., Shoham, Y. & Lamed, R. (1998). Cellulosomes—Structure and ultrastructure. *J Struct Biol*, *124*, 221-234.

Bjerre, A. B., Olesen, A. B., Fernqvist, T., Plöger, A. & Schmidt, A. S. (1996). Pretreatment of wheat straw using combined wet oxidation and alkaline hydrolysis resulting in convertible cellulose and hemicellulose. *Biotechnol Bioeng*, *49*, 568-577.

Blair, B. G. & Anderson, K. L. (1999). Regulation of cellulose-inducible structures of *Clostridium cellulovorans*. *Can J Microbiol*, *45*, 242-249.

Blouzard, J. C., Coutinho, P. M., Fierobe, H. P., Henrissat, B., Lignon, S., Tardif, C., Pagès, S. & de Philip, P. (2010). Modulation of cellulosome composition in *Clostridium cellulolyticum*: adaptation to the polysaccharide environment revealed by proteomic and carbohydrate-active enzyme analyses. *Proteomics*, *10*, 541-554.

Cann, I., Bernardi, R. C. & Mackie, R. I. (2016). Cellulose degradation in the human gut: *Ruminococcus champanellensis* expands the cellulosome paradigm. *Environ Microbiol*, *18*, 307-310.

Chen, W. H., Pen, B. L., Yu, C. T. & Hwang, W. S. (2011). Pretreatment efficiency and structural characterization of rice straw by an integrated process of dilute-acid and steam explosion for bioethanol production. *Bioresour Technol*, *102*, 2916-2924.

Cho, W., Jeon, S. D., Shim, H. J., Doi, R. H. & Han, S. O. (2010). Cellulosomic profiling produced by *Clostridium cellulovorans* during growth on different carbon sources explored by the cohesin marker. *J Biotechnol*, *145*, 233-239.

Cosgrove, D. J. (2005). Growth of the plant cell wall. *Nat Rev Mol Cell Biol*, *6*, 850-861.

Cumming, C. M., Rizkallah, H. D., McKendrick, K. A., Abdel-Massih, R. M., Baydoun, E. A. & Brett, C. T. (2005). Biosynthesis and cell-wall deposition of a pectin–xyloglucan complex in pea. *Planta*, *222*, 546-555.

Doi, R. H., Goldstein, M., Hashida, S., Park, J. S. & Takagi, M. (1994). The *Clostridium cellulovorans* cellulosome. *Crit Rev Microbiol*, *20*, 87-93.

Doi, R. H., Park, J. S., Liu, C. c., Malburg, Jr. L. M., Tamaru, Y., Ichiishi, A. & Ibrahim, A. (1998). Cellulosome and noncellulosomal cellulases of *Clostridium cellulovorans*. *Extremophiles*, *2*, 53-60.

Doi, R. H. & Tamaru, Y. (2001). The *Clostridium cellulovorans* cellulosome: an enzyme complex with plant cell wall degrading activity. *Chem Rec*, *1*, 24-32.

Eliuk, S. & Makarov, A. (2015). Evolution of Orbitrap mass spectrometry instrumentation. *Annu Rev Anal Chem*, *8*, 61-80.

Esaka, K., Aburaya, S., Morisaka, H., Kuroda, K. & Ueda, M. (2015). Exoproteome analysis of *Clostridium cellulovorans* in natural soft-biomass degradation. *AMB Express*, *5*, 1-8.

Gal, L., Pages, S., Gaudin, C., Belaich, A., Reverbel-Leroy, C., Tardif, C., & Belaich, J. P. (1997). Characterization of the cellulolytic complex (cellulosome) produced by *Clostridium cellulolyticum*. *Appl Environ Microbiol*, *63*, 903-909.

Garrote, G., Falqué, E., Domínguez, H. & Parajó, J. C. (2007). Autohydrolysis of agricultural residues: study of reaction byproducts. *Bioresour Technol*, *98*, 1951-1957.

Giddings, J. C. (1991). Unified separation science (Wiley New York, NY).

Girisuta, B., Dussan, K., Haverty, D., Leahy, J. J. & Hayes, M. H. B. (2013). A kinetic study of acid catalysed hydrolysis of sugar cane bagasse to levulinic acid. *Chem Eng J*, *217*, 61-70.

Groom, M. J., Gray, E. M. & Townsend, P. A. (2008). Biofuels and biodiversity: principles for creating better policies for biofuel production. *Conserv Biol*, *22*, 602-609.

Hayashi, T. (1989). Xyloglucans in the primary cell wall. *Annu Rev Plant Biol*, *40*, 139-168.

Hebert, A. S., Richards, A. L., Bailey, D. J., Ulbrich, A., Coughlin, E. E., Westphall, M. S. & Coon, J. J. (2014). The one hour yeast proteome. *Mol Cell Proteomics*, *13*, 339-347.

Iwasaki, M., Miwa, S., Ikegami, T., Tomita, M., Tanaka, N. & Ishihama, Y. (2010). One-dimensional capillary liquid chromatographic separation coupled with tandem mass spectrometry unveils the *Escherichia coli* proteome on a microarray scale. *Anal Chem*, *82*, 2616-2620.

Jones, B. & Nachtsheim, C. J. (2011). A class of three-level designs for definitive screening in the presence of second-order effects. *J Qual Technol*, *43*, 1.

Kalli, A. & Hess, S. (2012). Effect of mass spectrometric parameters on peptide and protein identification rates for shotgun proteomic experiments on an LTQ-orbitrap mass analyzer. *Proteomics*, *12*, 21-31.

Kobayashi, M., Matoh, T. & Azuma, J. (1996). Two chains of rhamnogalacturonan II are cross-linked by borate-diol ester bonds in higher plant cell walls. *Plant Physiol*, *110*, 1017-1020.

Kosugi, A., Murashima, K. & Doi, R. H. (2002). Xylanase and acetyl xylan esterase activities of XynA, a key subunit of the *Clostridium cellulovorans* cellulosome for xylan degradation. *Appl Environ Microbiol*, *68*, 6399-6402.

Kristensen, J. B., Thygesen, L. G., Felby, C., Jørgensen, H. & Elder, T. (2008). Cell-wall structural changes in wheat straw pretreated for bioethanol production. *Biotechnol Biofuels*, *1*, 5.

Lee, J. (1997). Biological conversion of lignocellulosic biomass to ethanol. *J Biotechnol*, *56*, 1-24.

Lemaire, M., Ohayon, H., Gounon, P., Fujino, T. & Béguin, P. (1995). OlpB, a new outer layer protein of *Clostridium thermocellum*, and binding of its S-layer-like domains to components of the cell envelope. *J Bacteriol*, *177*, 2451-2459.

Li, C., Knierim, B., Manisseri, C., Arora, R., Scheller, H. V., Auer, M., Vogel, K. P., Simmons, B. A. & Singh, S. (2010a). Comparison of dilute acid and ionic liquid pretreatment of switchgrass: Biomass recalcitrance, delignification and enzymatic saccharification. *Bioresour Technol*, *101*, 4900-4906.

Li, M. F., Fan, Y. M., Xu, F., Sun, R. C. & Zhang, X. L. (2010b). Cold sodium hydroxide/urea based pretreatment of bamboo for bioethanol production: Characterization of the cellulose rich fraction. *Ind Crops Prod*, *32*, 551-559.

Li, X., Kim, T. H. & Nghiem, N. P. (2010c). Bioethanol production from corn stover using aqueous ammonia pretreatment and two-phase simultaneous saccharification and fermentation (TPSSF). *Bioresour Technol*, *101*, 5910-5916.

Lynd, L. R., Wyman, C. E. & Gerngross, T. U. (1999). Biocommodity engineering. *Biotechnol Prog*, *15*, 777-793.

Lynd, L. R., Zyl, W. H. v., McBride, J. E. & Laser, M. (2005). Consolidated bioprocessing of cellulosic biomass: an update. *Curr Opin Biotechnol*, *16*, 577-583.

Matsui, K., Bae, J., Esaka, K., Morisaka, H., Kuroda, K. & Ueda, M. (2013). Exoproteome profiles of *Clostridium cellulovorans* grown on various carbon sources. *Appl Environ Microbiol*, *79*, 6576-6584.

Menon, V. & Rao, M. (2012). Trends in bioconversion of lignocellulose: biofuels, platform chemicals & biorefinery concept. *Prog Energy Combust Sci*, *38*, 522-550.

Michalski, A., Cox, J. & Mann, M. (2011). More than 100,000 detectable peptide species elute in single shotgun proteomics runs but the majority is inaccessible to data-dependent LC−MS/MS. *J Proteome Res*, *10*, 1785-1793.

Mohand-Oussaid, O., Payot, S., Guedon, E., Gelhaye, E., Youyou, A. & Petitdemange, H. (1999). The extracellular xylan degradative system in *Clostridium cellulolyticum* cultivated on xylan: evidence for cell-free cellulosome production. *J Bacteriol*, *181*, 4035-4040.

Mohnen, D. (2008). Pectin structure and biosynthesis. *Curr Opin Plant Biol*, *11*, 266-277.

Mood, S. H., Golfeshan, A. H., Tabatabaei, M., Jouzani, G. S., Najafi, G. H., Gholami, M. & Ardjmand, M. (2013). Lignocellulosic biomass to bioethanol, a comprehensive review with a focus on pretreatment. *Renew Sust Energ Rev*, *27*, 77-93.

Morisaka, H., Matsui, K., Tatsukami, Y., Kuroda, K., Miyake, H., Tamaru, Y. & Ueda, M. (2012). Profile of native cellulosomal proteins of *Clostridium cellulovorans* adapted to various carbon sources. *AMB Express*, *2*, 37.

Murashima, K., Kosugi, A. & Doi, R. H. (2003). Synergistic effects of cellulosomal xylanase and cellulases from Clostridium cellulovorans on plant cell wall degradation. *J Bacteriol*, *185*, 1518-1524.

Nataf, Y., Bahari, L., Kahel-Raifer, H., Borovok, I., Lamed, R., Bayer, E. A., Sonenshein, A. L. & Shoham, Y. (2010). *Clostridium thermocellum* cellulosomal genes are regulated by extracytoplasmic polysaccharides via alternative sigma factors. *Proc Natl Acad Sci U S A*, *107*, 18646-18651.

Ng, T., Weimer, P. & Zeikus, J. (1977). Cellulolytic and physiological properties of *Clostridium thermocellum. Archives Microbiol*, *114*, 1-7.

O'Neill, M., Albersheim, P. & Darvill, A. (1990). 12 - The Pectic Polysaccharides of Primary Cell Walls. In *Methods in Plant Biochemistry*, P. M. Dey, ed. (Academic Press), pp. 415-441.

Ohara, H., Karita, S., Kimura, T., Sakka, K. & Ohmiya, K. (2000). Characterization of the cellulolytic complex (cellulosome) from *Ruminococcus albus. Biosci Biotechnol Biochem*, *64*, 254-260.

Ridley, B. L., O'Neill, M. A. & Mohnen, D. (2001). Pectins: structure, biosynthesis, and oligogalacturonide-related signaling. *Phytochemistry*, *57*, 929-967.

Rydzak, T., McQueen, P. D., Krokhin, O. V., Spicer, V., Ezzati, P., Dwivedi, R. C., Shamshurin, D., Levin, D. B., Wilkins, J. A. & Sparling, R. (2012). Proteomic analysis of *Clostridium thermocellum* core metabolism: relative protein expression profiles and growth phase-dependent changes in protein expression. *BMC Microbiol*, *12*, 214.

Saha, B. C. (2003). Hemicellulose bioconversion. *J Ind Microbiol Biotechnol*, *30*, 279-291.

Shoham, Y., Lamed, R. & Bayer, E. A. (1999). The cellulosome concept as an efficient microbial strategy for the degradation of insoluble polysaccharides. *Trends Microbiol*, *7*, 275-281.

Sleat, R., Mah, R. A. & Robinson, R. (1984). Isolation and characterization of an anaerobic, cellulolytic bacterium, *Clostridium cellulovorans* sp. nov. *Appl Environ Microbiol*, *48*, 88-93.

Sun, Y. & Cheng, J. (2002). Hydrolysis of lignocellulosic materials for ethanol production: a review. *Bioresour Technol*, *83*, 1-11.

Tamaru, Y. & Doi, R. H. (2001). Pectate lyase A, an enzymatic subunit of the *Clostridium cellulovorans* cellulosome. *Proc Natl Acad Sci U S A*, *98*, 4125-4129.

Tamaru, Y., Miyake, H., Kuroda, K., Nakanishi, A., Matsushima, C., Doi, R. H. & Ueda, M. (2011). Comparison of the mesophilic cellulosome-producing *Clostridium cellulovorans* genome with other cellulosome-related clostridial genomes. *Microb Biotechnol*, *4*, 64-73.

Tamaru, Y., Miyake, H., Kuroda, K., Ueda, M. & Doi, R. H. (2010). Comparative genomics of the mesophilic cellulosome-producing *Clostridium cellulovorans* and its application to biofuel production via consolidated bioprocessing. *Environment Technol*, *31*, 889-903.

Tanaka, N., Kobayashi, H., Ishizuka, N., Minakuchi, H., Nakanishi, K., Hosoya, K. & Ikegami, T. (2002). Monolithic silica columns for high-efficiency chromatographic separations. *J Chromatogr A*, *965*, 35-49.

Vincken, J. P., Schols, H. A., Oomen, R. J., McCann, M. C., Ulvskov, P., Voragen, A. G. & Visser, R. G. (2003). If homogalacturonan were a side chain of rhamnogalacturonan I. Implications for cell wall architecture. *Plant Physiol*, *132*, 1781-1789.

Vollmer, M., Hörth, P. & Nägele, E. (2004). Optimization of two-dimensional off-line LC/MS separations to improve resolution of complex proteomic samples. *Anal Chem*, *76*, 5180-5185.

Willats, W. G., McCartney, L., Mackie, W. & Knox, J. P. (2001). Pectin: cell biology and prospects for functional analysis. *Plant Mol Biol*, *47*, 9-27.

Xu, C., Huang, R., Teng, L., Wang, D., Hemme, C. L., Borovok, I., He, Q., Lamed, R., Bayer, E. A. & Zhou, J. (2013). Structure and regulation of the cellulose degradome in *Clostridium cellulolyticum*. *Biotechnol Biofuels*, *6*, 73.

Xu, P., Duong, D. M. & Peng, J. (2009). Systematical optimization of reverse-phase chromatography for shotgun proteomics. *J Proteome Res*, *8*, 3944-3950.

Xu, Q., Gao, W., Ding, S. Y., Kenig, R., Shoham, Y., Bayer, E. A. & Lamed, R. (2003). The cellulosome system of *Acetivibrio cellulolyticus* includes a novel type of adaptor protein and a cell surface anchoring protein. *J Bacteriol*, *185*, 4548-4557.

Xu, Q., Resch, M. G., Podkaminer, K., Yang, S., Baker, J. O., Donohoe, B. S., Wilson, C., Klingeman, D. M., Olson, D. G. & Decker, S. R. (2016). Dramatic performance of *Clostridium thermocellum* explained by its wide range of cellulase modalities. *Sci Adv*, *2*, e1501254.

Zhang, Y. H. P. & Lynd, L. R. (2005). Cellulose utilization by *Clostridium thermocellum*: bioenergetics and hydrolysis product assimilation. *Proc Natl Acad Sci U S A*, *102*, 7321-7325.

In: Proteomics
Editor: Ricardo Parker
ISBN: 978-1-53616-440-4

Chapter 5

STANDARDIZATION AND QUANTIFICATION BY COMPARATIVE FLUORESCENCE GEL ELECTROPHORESIS (COFGE)

Alina Nippes[1], Doreen Ackermann[1] and Simone König[1,*]
[1]Core Unit Proteomics, Interdisciplinary Center for Clinical Research,
Westfalian Wilhelms-University of Münster,
Münster, Germany

ABSTRACT

Comparative fluorescence gel electrophoresis (CoFGE) was invented in 2009 and has since then become the prime method for reproducible coordinate assignment in two-dimensional protein polyacrylamide gel electrophoresis. The method is based on the use of at least two fluorescent dyes in one gel run: one for the analyte and one for an internal protein marker, which generates a net of reference spots across the gel. An additional dye can be used to label a standard dilution. Although the marker originally was intended solely for the purpose of providing anchor spots for the correction of analyte coordinates, it may be additionally used for

* Corresponding Author's E-mail: koenigs@uni-muenster.de.

quantification as the fluorescence intensity can be related to the protein amount causing it. The reference net is formed by 8 - 10 proteins ranging from 8 - 100 kDa and 40 wells for marker application so that several standard concentrations can be applied in one run. The major drawbacks of gel electrophoresis – reproducibility and lack of standardisation – thus finally vanish.

Keywords: 1D-PAGE, 2D-PAGE, protein analysis, quantification

COMMENTARY

Comparative fluorescence gel electrophoresis (CoFGE [1-4]) set out to battle the problem of low reproducibility in two-dimensional polyacrylamide gel electrophoresis (2D-PAGE). First developed for vertical PAGE [1, 2], it works at its best in horizontal electrophoresis [3]. The method uses multifluorescent labelling to cover the 2D-gel with a net of known marker protein spots overlaying the analyte protein.

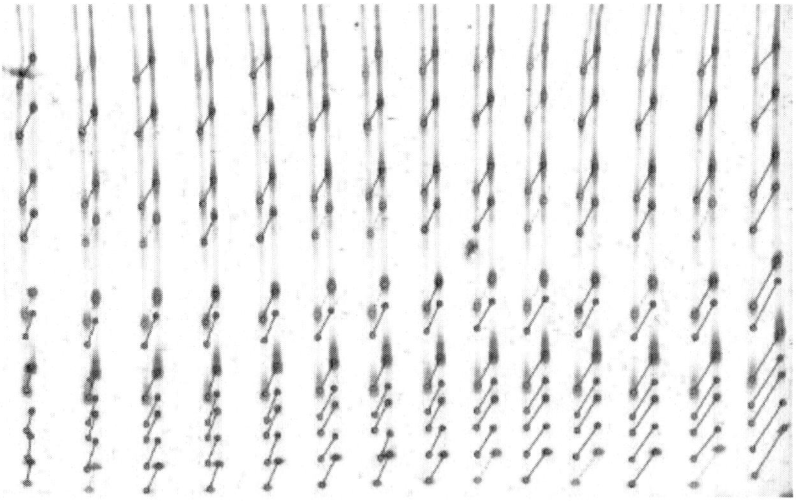

Figure 1. Horizontal CoFGE illustrating the principle of coordinate correction. Reference spots are created across the gel (choice of 8-10 proteins, 14 wells on commercial Mercator gels (Serva)) which are related (warped) to a predefined master grid. Once the reference on the analyte gel has been corrected, the match vectors are applied to the analyte spots [5].

It allows improved matching of different gels – as long as they have been generated with the same protocol – by correcting them with the reference grid (Figure 1).

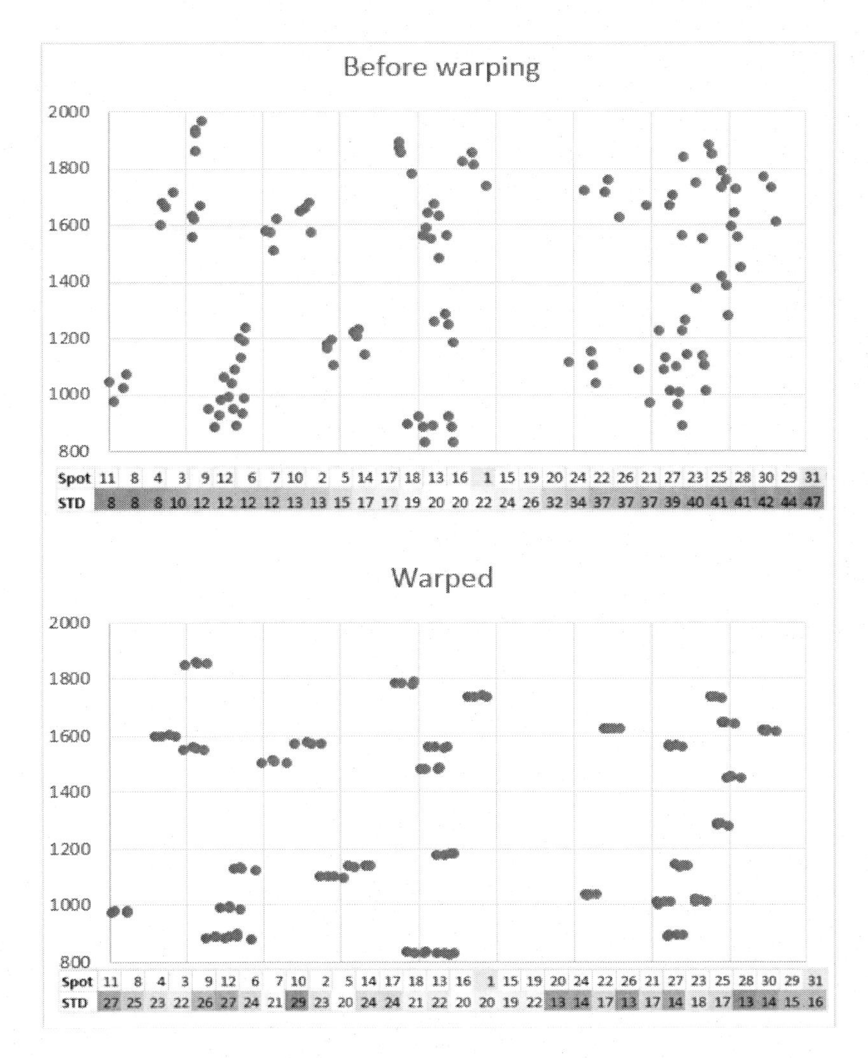

Figure 2. Exemplary spots determined on four Escherichia coli gels before and after warping visualized in the Delta2D coordinate grid (with spot# and standard deviation of x-coordinate). Most improvement is seen for the y-coordinate (molecular weight dimension) as the marker experiences the same progression through the gel as the analyte during the run. The standard deviations for the x-dimension shown beneath the coordinate system improve particularly for the spots on the basic side of the gel. Results can be further improved by introducing reference markers for the pI [4].

Improvements in coordinate assignment are primarily achieved for the molecular weigth dimension as marker proteins co-run with the sample and experience the same gel distortions (Figure 2). Larger, but for most purposes acceptable, deviation occurs in the p*I*-direction. Azo dyes can be used for the correction of the first dimension, if necessary [4]. CoFGE reduces the error in coordinate assignment of protein spots to well below 10%, finally enabling true sample comparison across laboratories and thus 2D-map databases.

Since its invention in 2009, the method has been commercialized and become much more user-friendly; software has taken over image analyses procedures which were formerly performed manually [5].

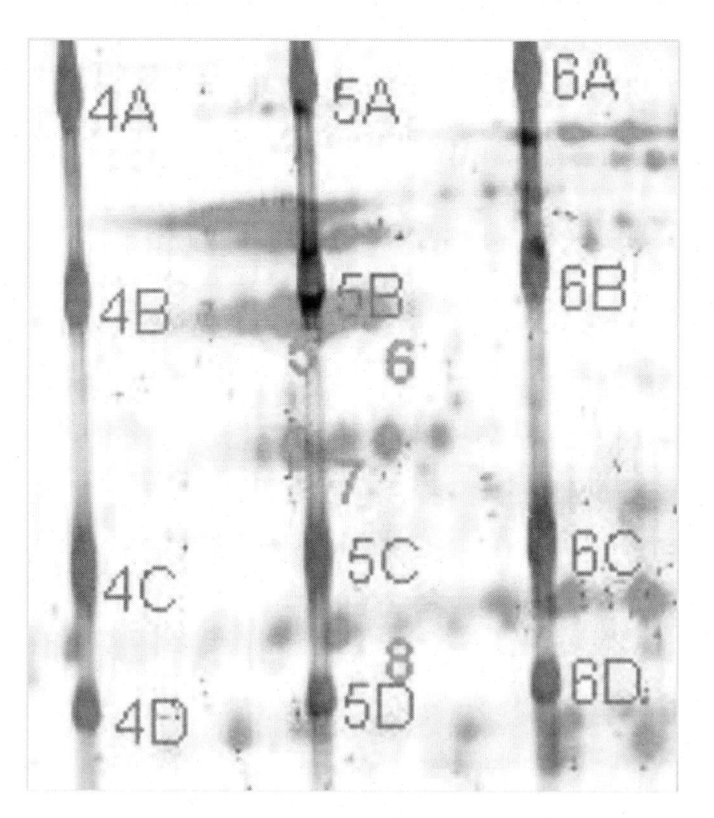

Figure 3. Zoom into an exemplary gel area showing *E. coli* proteins (spots 5-8) and marker spots 4A-6D. The molecular weight of protein 5 is about the same as that of the marker 5B, the intensity of 5B allows an estimation of the amount of protein present in spots 5 or 6. For spots farther away from reference spots, rule of proportions applies.

CoFGE was much improved with respect to handling and reproducibility in horizontal electrophoresis [3] and, meanwhile, pre-cast gels have become available that contain 14 slots for the protein marker grid in addition to the well for the p*I*-strip. However, CoFGE can do much more. As an internal standard with known concentration is applied for correct coordinate output, it may be related to its fluorescence intensity and thus be used to estimate the molecular weight and the protein amount on analyte spots of interest (Figure 3). Obviously, differences due to the specific properties of individual dyes need to be determined beforehand, but in case of well matched dyes such as Cy3/Cy5 (GE) or Sci3/Sci5 (Serva) they may be minimal.

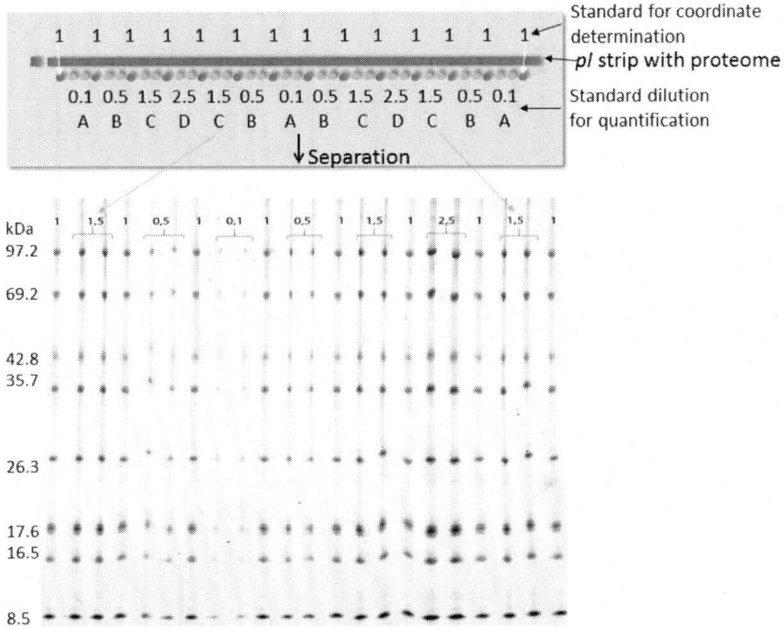

Standard for coordinate determination
p*I* strip with proteome
Standard dilution for quantification
1 1 1 1 1 1 1 1 1 1 1 1 1 1
0.1 0.5 1.5 2.5 1.5 0.5 0.1 0.5 1.5 2.5 1.5 0.5 0.1
A B C D C B A B C D C B A
Separation
kDa
97.2
69.2
42.8
35.7
26.3
17.6
16.5
8.5

Figure 4. Approach to add a marker dilution series to a CoFGE gel. Shown at the top is a schematic of a commercial Mercator gel (Serva), which contains 14 wells for the application of marker (protein standard mix with predefined concentration – set to 1 here). Two more slots can be punched manually between each pre-formed well to hold marker of differing concentrations. Multiples of 1 (A 0.1, B 0.5, C 1.5, D 2.5) were used. Differences of the gel matrix were taken into account by the application of an ABCDCBABCDCBA pattern across the gel (two spots for each concentration). Below: Zoom into resulting gel.

On a commercial CoFGE-gel containing 14 wells for the marker, two more wells can be punched manually between each two without loss of resolution (Figure 4) leading to 26 more wells, which can be filled with marker of other concentrations. A dilution series of 0.1 to 2.5 of a selected protein marker mix (concentration normalized to 1, Figure 4) applied in an alternating pattern to acknowledge changes in the gel matrix across the gel, provides good calibration and reference for the analysis of unknowns. The read-back error for marker spots is in the order of 10% and better with low concentrations exhibiting the highest errors.

With prices for dyes and detection technology dropping, fluorescent labelling becomes affordable for routine use. Performing a CoFGE experiment has become much easier due to commercialisation of consumables and advances in software tools. The major drawback of gel electrophoresis – reproducibility – finally vanishes.

ACKNOWLEDGMENTS

The authors thank B. Müller of Serva Electrophoresis and G. Wrettos of Decodon for helpful discussions.

REFERENCES

[1] König, S., W. Wang, L. Grün, D. Ackermann. Improved gel electrophoresis. *WWU 2009*, EP11167383.6, 2011; 12729346.2-1554, 2012, EP 2 715 331.

[2] Ackermann, D., Wang, W., Streipert, B., Geib, B., Grün, L., König, S. Comparative fluorescence two-dimensional gel electrophoresis using a gel strip sandwich assembly for the simultaneous on-gel generation of a reference protein spot grid. *Electrophoresis,* 2012, 33:1406.

[3] Hanneken, M., König, S. Horizontal comparative fluorescence two-dimensional gel electrophoresis (hCoFGE) for improved spot coordinate detection. *Electrophoresis,* 2014, 35: 1118.

[4] Hanneken, M., Slais, K., König, S. pI control in comparative fluorescence two-dimensional gel electrophoresis (CoFGE) using amphoteric azo dyes. *EuPa Open Proteomics*, 2015, 8:36.

[5] Nippes, A., Ackermann, D., König, S. Analysis of CoFGE experiments with Delta2D. *Mercator J. Biomol. Anal.,* 2018, 2:3, urn:nbn:de:hbz:6-97129496060.

INDEX

#

A

B

C

D

E

F

G

H

I

L

M

S

T

U

V

W

Related Nova Publications

The Mechanics of Life: A Closer Look at the Inner Workings of Nature

Author: Timothy Ganesan, Ph.D.

Series: Systems Biology – Theory, Techniques and Applications

Book Description: This book is devoted to current ideas and developments in the biological sciences which stretch into fields such as engineering, medicine, quantum physics, computer modeling and genetics.

Hardcover ISBN: 978-1-53612-937-3
Retail Price: $195

Functional Genomics: Novel Insights, Applications and Future Challenges

Editors: Holger Uffe and Olaf Philip

Series: Systems Biology – Theory, Techniques and Applications

Book Description: In Chapter One, the authors discuss the gene silencing tool RNA interference, or RNAi, which functions at both transcriptional and post-transcriptional levels. In Chapter Two, the authors study the origins of genome engineering, including its adaptation from prokaryotic to eukaryotic systems and applications.

Softcover ISBN: 978-1-53612-565-8
Retail Price: $82

To see a complete list of Nova publications, please visit our website at www.novapublishers.com

Related Nova Publications

Systems Synthetic Biology: System Models, User-Oriented Specifications, and Applications

Authors: Bor-Sen Chen and Chih-Yuan Hsu

Series: Systems Biology – Theory, Techniques and Applications

Book Description: In this book, the synthetic gene circuits are modeled by nonlinear stochastic systems to consider random genetic variations and random in vivo environmental disturbances. The authors' design purpose is to engineer a robust genetic circuit to achieve a desired behavior or product to tolerate intrinsic random fluctuation and environmental disturbance in the host cell.

Hardcover ISBN: 978-1-53610-210-9
Retail Price: $195

Biological Energetics

Author: Pang Xiao Feng

Series: Systems Biology – Theory, Techniques and Applications

Book Description: This book presents a complete review of biological energetics and responds systematically that there is no possible activity without bio-energy in the living bodies.

Hardcover ISBN: 978-1-63484-811-4
Retail Price: $270

To see a complete list of Nova publications, please visit our website at www.novapublishers.com